humble math - 100 days of timed tests

The book belongs to:

_____

—·—··—·—··—·—··—·—·· —

_____

Day 1

Name:...........................

Score: /20

Time: ....

0 x 6	0 x 5	0 x 4	0 x 3	0 x 2	0 x 1	0 x 0
3 x 0	2 x 0	1 x 0	0 x 0	0 x 9	0 x 8	0 x 7
1 x 1	9 x 0	8 x 0	7 x 0	6 x 0	5 x 0	4 x 0
1 x 8	1 x 7	1 x 6	1 x 5	1 x 4	1 x 3	1 x 2
5 x 1	4 x 1	3 x 1	2 x 1	1 x 1	0 x 1	1 x 9
2 x 2	2 x 1	2 x 0	9 x 1	8 x 1	7 x 1	6 x 1
2 x 9	2 x 8	2 x 7	2 x 6	2 x 5	2 x 4	2 x 3
6 x 2	5 x 2	4 x 2	3 x 2	2 x 2	1 x 2	0 x 2
3 x 3	3 x 2	3 x 1	3 x 0	9 x 2	8 x 2	7 x 2

Day 2

Name:..........................

Score: /20

Time: ....

3 x 4	3 x 5	3 x 6	3 x 7	3 x 8	3 x 9	0 x 3
1 x 3	2 x 3	3 x 3	4 x 3	5 x 3	6 x 3	7 x 3
8 x 3	9 x 3	4 x 0	4 x 1	4 x 2	4 x 3	4 x 4
4 x 5	4 x 6	4 x 7	4 x 8	4 x 9	0 x 4	1 x 4
2 x 4	3 x 4	4 x 4	5 x 4	6 x 4	7 x 4	8 x 4
9 x 4	5 x 0	1 x 5	2 x 5	3 x 5	4 x 5	5 x 5
6 x 5	7 x 5	8 x 5	9 x 5	5 x 0	5 x 1	5 x 2
5 x 3	5 x 4	5 x 5	5 x 6	5 x 7	5 x 8	5 x 9
6 x 0	6 x 1	6 x 2	6 x 3	6 x 4	6 x 5	6 x 6

Name:........................

Score: /20

Time: ....

3 x 6	2 x 6	1 x 6	0 x 6	6 x 9	6 x 8	6 x 7
7 x 0	9 x 6	8 x 6	7 x 6	6 x 6	5 x 6	4 x 6
7 x 7	7 x 6	7 x 5	7 x 4	7 x 3	7 x 2	7 x 1
4 x 7	3 x 7	2 x 7	1 x 7	0 x 7	7 x 9	7 x 8
8 x 1	8 x 0	9 x 7	8 x 7	7 x 7	6 x 7	5 x 7
8 x 8	8 x 7	8 x 6	8 x 5	8 x 4	8 x 3	8 x 2
5 x 8	4 x 8	3 x 8	2 x 8	1 x 8	0 x 8	8 x 9
9 x 2	9 x 1	9 x 0	9 x 8	8 x 8	7 x 8	6 x 8
9 x 9	9 x 8	9 x 7	9 x 6	9 x 5	9 x 4	9 x 3

Name:...........................

Score: /20

Time: ....

0 x 9	1 x 9	2 x 9	3 x 9	4 x 9	5 x 9	6 x 9
7 x 9	8 x 9	9 x 9	1 x 2	2 x 3	3 x 4	4 x 5
5 x 6	6 x 7	7 x 8	8 x 9	9 x 1	2 x 8	3 x 7
4 x 6	9 x 0	0 x 2	2 x 4	8 x 1	8 x 6	5 x 2
4 x 0	9 x 1	5 x 3	1 x 3	1 x 6	7 x 2	5 x 1
4 x 1	4 x 2	8 x 3	9 x 3	4 x 0	4 x 3	4 x 4
4 x 8	4 x 9	4 x 5	4 x 6	4 x 7	0 x 4	1 x 4
5 x 4	6 x 4	2 x 4	3 x 4	4 x 4	7 x 4	8 x 4
2 x 5	3 x 5	9 x 4	5 x 0	1 x 5	4 x 5	5 x 5

Day 5

Name:.........................

Score: /20

Time: ....

1 x 1	9 x 0	8 x 0	0 x 3	0 x 2	0 x 1	0 x 0
1 x 8	1 x 7	1 x 6	0 x 0	0 x 9	0 x 8	0 x 7
5 x 1	4 x 1	3 x 1	7 x 0	6 x 0	5 x 0	4 x 0
2 x 2	2 x 1	2 x 0	1 x 5	1 x 4	1 x 3	1 x 2
2 x 1	1 x 1	0 x 1	1 x 9	0 x 6	0 x 5	0 x 4
9 x 1	8 x 1	7 x 1	6 x 1	3 x 0	2 x 0	1 x 0
2 x 9	2 x 8	2 x 7	2 x 6	2 x 5	2 x 4	2 x 3
6 x 2	5 x 2	4 x 2	3 x 2	2 x 2	1 x 2	0 x 2
3 x 3	3 x 2	3 x 1	3 x 0	9 x 2	8 x 2	7 x 2

Day 6

Name:..........................

Score: /20

Time: ....

2	3	1	4	3	3	0
x 4	x 4	x 6	x 7	x 8	x 0	x 3

1	2	3	4	5	6	7
x 1	x 3	x 3	x 3	x 3	x 3	x 3

8	9	4	4	2	4	4
x 3	x 9	x 0	x 1	x 2	x 3	x 4

4	9	4	4	7	0	1
x 5	x 6	x 7	x 0	x 9	x 4	x 4

2	3	4	5	6	7	8
x 4	x 4	x 4	x 1	x 6	x 4	x 0

7						
9	5	1	2	3	4	5
x 4	x 0	x 5	x 7	x 5	x 4	x 5

6	7	8	9	5	6	5
x 5	x 9	x 5	x 7	x 0	x 1	x 2

2	5	5	1	5	5	5
x 3	x 4	x 5	x 6	x 7	x 8	x 9

6	5	6	7	6	8	6
x 0	x 1	x 2	x 3	x 4	x 5	x 6

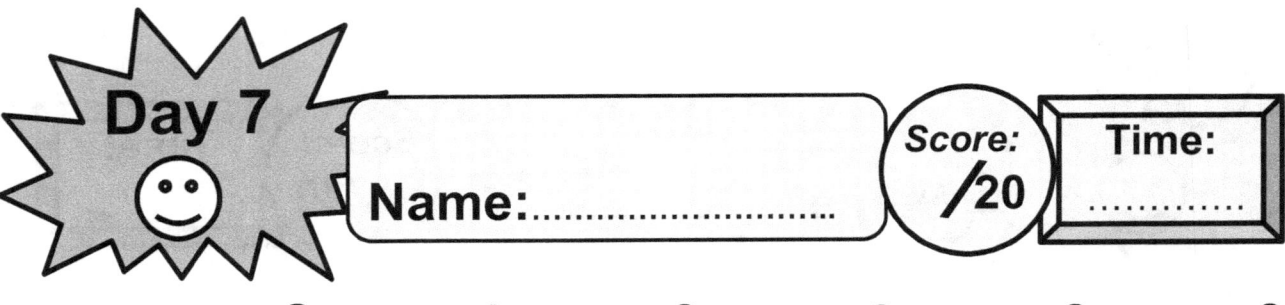

3 x 0	2 x 6	1 x 9	0 x 6	6 x 8	6 x 7	6 x 7
7 x 0	9 x 1	8 x 6	7 x 2	6 x 6	5 x 3	4 x 6
5 x 2	4 x 6	6 x 5	7 x 4	8 x 3	7 x 2	9 x 1
4 x 0	3 x 7	2 x 9	1 x 7	0 x 2	7 x 9	7 x 8
8 x 1	8 x 0	9 x 7	8 x 7	7 x 7	6 x 7	5 x 7
8 x 8	5 x 7	8 x 6	2 x 5	8 x 4	9 x 3	8 x 2
6 x 8	4 x 5	3 x 8	2 x 4	1 x 8	0 x 8	8 x 9
9 x 2	9 x 1	9 x 0	9 x 8	8 x 8	7 x 8	6 x 8
9 x 9	9 x 8	9 x 7	9 x 6	9 x 5	9 x 4	9 x 3

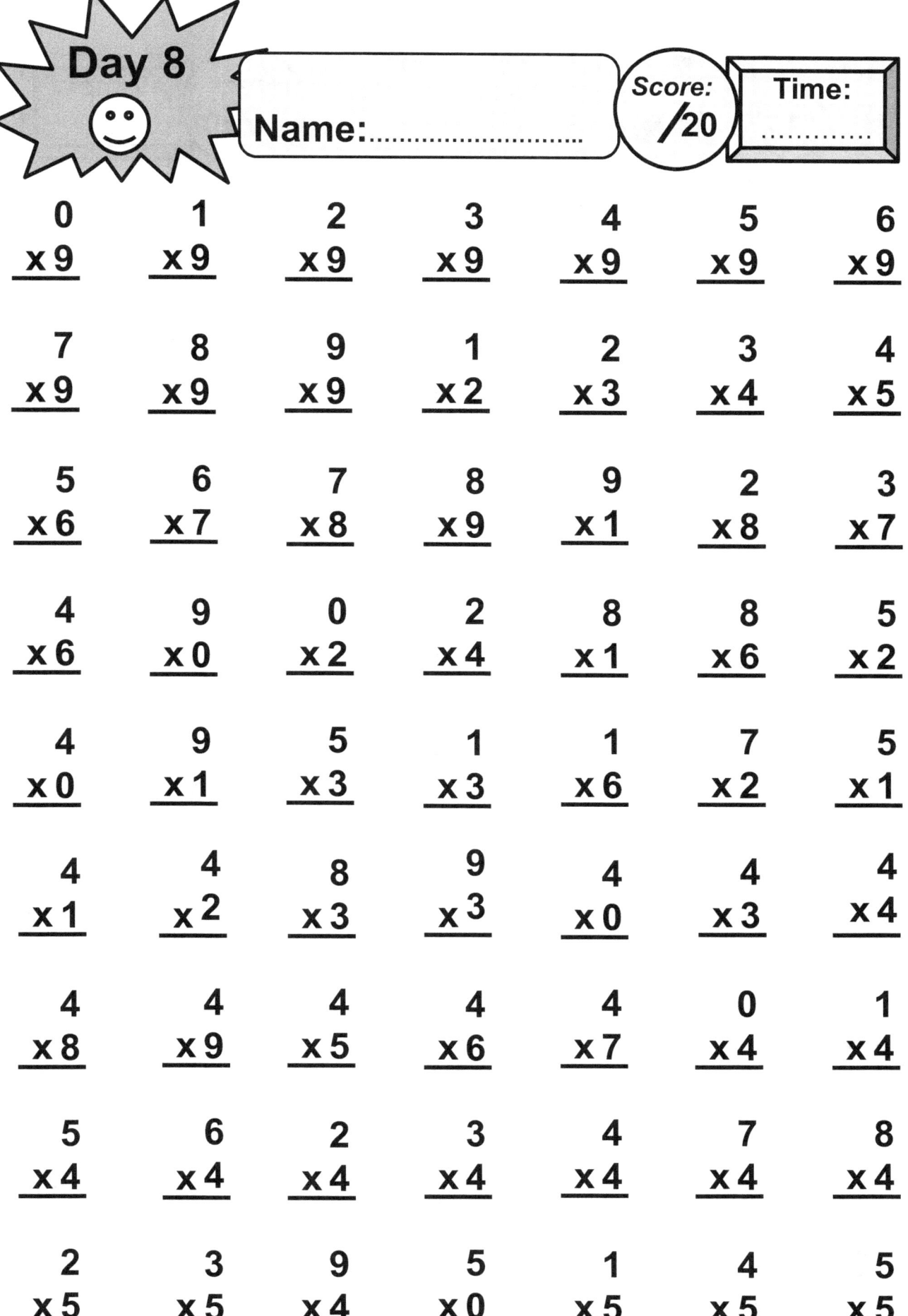

Day 8

Name:............................

Score: /20

Time: ....

0	1	2	3	4	5	6
x 9	x 9	x 9	x 9	x 9	x 9	x 9

7	8	9	1	2	3	4
x 9	x 9	x 9	x 2	x 3	x 4	x 5

5	6	7	8	9	2	3
x 6	x 7	x 8	x 9	x 1	x 8	x 7

4	9	0	2	8	8	5
x 6	x 0	x 2	x 4	x 1	x 6	x 2

4	9	5	1	1	7	5
x 0	x 1	x 3	x 3	x 6	x 2	x 1

4	4	8	9	4	4	4
x 1	x 2	x 3	x 3	x 0	x 3	x 4

4	4	4	4	4	0	1
x 8	x 9	x 5	x 6	x 7	x 4	x 4

5	6	2	3	4	7	8
x 4	x 4	x 4	x 4	x 4	x 4	x 4

2	3	9	5	1	4	5
x 5	x 5	x 4	x 0	x 5	x 5	x 5

Name:...........................

Score: /20

Time: ....

0	0	0	0	0	0	0
x 6	x 5	x 4	x 3	x 2	x 1	x 0

3	2	1	0	0	0	0
x 0	x 0	x 0	x 0	x 9	x 8	x 7

1	9	8	7	6	5	4
x 1	x 0	x 0	x 0	x 0	x 0	x 0

1	1	1	1	1	1	1
x 8	x 7	x 6	x 5	x 4	x 3	x 2

5	4	3	2	1	0	1
x 1	x 1	x 1	x 1	x 1	x 1	x 9

2	2	2	9	8	7	6
x 2	x 1	x 0	x 1	x 1	x 1	x 1

2	2	2	2	2	2	2
x 9	x 8	x 7	x 6	x 5	x 4	x 3

6	5	4	3	2	1	0
x 2	x 2	x 2	x 2	x 2	x 2	x 2

3	3	3	3	9	8	7
x 3	x 2	x 1	x 0	x 2	x 2	x 2

3 x 4	3 x 5	3 x 6	3 x 7	3 x 8	3 x 9	0 x 3
1 x 3	2 x 3	3 x 3	4 x 3	5 x 3	6 x 3	7 x 3
8 x 3	9 x 3	4 x 0	4 x 1	4 x 2	4 x 3	4 x 4
4 x 5	4 x 6	4 x 7	4 x 8	4 x 9	0 x 4	1 x 4
2 x 4	3 x 4	4 x 4	5 x 4	6 x 4	7 x 4	8 x 4
9 x 4	5 x 0	1 x 5	2 x 5	3 x 5	4 x 5	5 x 5
6 x 5	7 x 5	8 x 5	9 x 5	5 x 0	5 x 1	5 x 2
5 x 3	5 x 4	5 x 5	5 x 6	5 x 7	5 x 8	5 x 9
6 x 0	6 x 1	6 x 2	6 x 3	6 x 4	6 x 5	6 x 6

3 x 6	2 x 6	1 x 6	0 x 6	6 x 9	6 x 8	6 x 7
7 x 0	9 x 6	8 x 6	7 x 6	6 x 6	5 x 6	4 x 6
7 x 7	7 x 6	7 x 5	7 x 4	7 x 3	7 x 2	7 x 1
4 x 7	3 x 7	2 x 7	1 x 7	0 x 7	7 x 9	7 x 8
8 x 1	8 x 0	9 x 7	8 x 7	7 x 7	6 x 7	5 x 7
8 x 8	8 x 7	8 x 6	8 x 5	8 x 4	8 x 3	8 x 2
5 x 8	4 x 8	3 x 8	2 x 8	1 x 8	0 x 8	8 x 9
9 x 2	9 x 1	9 x 0	9 x 8	8 x 8	7 x 8	6 x 8
9 x 9	9 x 8	9 x 7	9 x 6	9 x 5	9 x 4	9 x 3

Name:............................

Score: /20

Time: ....

0 x 9	1 x 9	2 x 9	3 x 9	4 x 9	5 x 9	6 x 9
7 x 9	8 x 9	9 x 9	1 x 2	2 x 3	3 x 4	4 x 5
5 x 6	6 x 7	7 x 8	8 x 9	9 x 1	2 x 8	3 x 7
4 x 6	9 x 0	0 x 2	2 x 4	8 x 1	8 x 6	5 x 2
4 x 0	9 x 1	5 x 3	1 x 3	1 x 6	7 x 2	5 x 1
4 x 1	4 x 2	8 x 3	9 x 3	4 x 0	4 x 3	4 x 4
4 x 8	4 x 9	4 x 5	4 x 6	4 x 7	0 x 4	1 x 4
5 x 4	6 x 4	2 x 4	3 x 4	4 x 4	7 x 4	8 x 4
2 x 5	3 x 5	9 x 4	5 x 0	1 x 5	4 x 5	5 x 5

Name:..........................

Score: /20

Time: ....

1 x 1	9 x 0	8 x 0	0 x 3	0 x 2	0 x 1	0 x 0
1 x 8	1 x 7	1 x 6	0 x 0	0 x 9	0 x 8	0 x 7
5 x 1	4 x 1	3 x 1	7 x 0	6 x 0	5 x 0	4 x 0
2 x 2	2 x 1	2 x 0	1 x 5	1 x 4	1 x 3	1 x 2
2 x 1	1 x 1	0 x 1	1 x 9	0 x 6	0 x 5	0 x 4
9 x 1	8 x 1	7 x 1	6 x 1	3 x 0	2 x 0	1 x 0
2 x 9	2 x 8	2 x 7	2 x 6	2 x 5	2 x 4	2 x 3
6 x 2	5 x 2	4 x 2	3 x 2	2 x 2	1 x 2	0 x 2
3 x 3	3 x 2	3 x 1	3 x 0	9 x 2	8 x 2	7 x 2

Name:...........................

Score: /20

Time: ....

2	3	1	4	3	3	0
x 4	x 4	x 6	x 7	x 8	x 0	x 3

1	2	3	4	5	6	7
x 1	x 3	x 3	x 3	x 3	x 3	x 3

8	9	4	4	2	4	4
x 3	x 9	x 0	x 1	x 2	x 3	x 4

4	9	4	4	7	0	1
x 5	x 6	x 7	x 0	x 9	x 4	x 4

2	3	4	5	6	7	8
x 4	x 4	x 4	x 1	x 6	x 4	x 0

7						
9	5	1	2	3	4	5
x 4	x 0	x 5	x 7	x 5	x 4	x 5

6	7	8	9	5	6	5
x 5	x 9	x 5	x 7	x 0	x 1	x 2

2	5	5	1	5	5	5
x 3	x 4	x 5	x 6	x 7	x 8	x 9

6	5	6	7	6	8	6
x 0	x 1	x 2	x 3	x 4	x 5	x 6

Name:................................

Score: /20

Time: ....

3 x 0	2 x 6	1 x 9	0 x 6	6 x 8	6 x 7	6 x 7
7 x 0	9 x 1	8 x 6	7 x 2	6 x 6	5 x 3	4 x 6
5 x 2	4 x 6	6 x 5	7 x 4	8 x 3	7 x 2	9 x 1
4 x 0	3 x 7	2 x 9	1 x 7	0 x 2	7 x 9	7 x 8
8 x 1	8 x 0	9 x 7	8 x 7	7 x 7	6 x 7	5 x 7
8 x 8	5 x 7	8 x 6	2 x 5	8 x 4	9 x 3	8 x 2
6 x 8	4 x 5	3 x 8	2 x 4	1 x 8	0 x 8	8 x 9
9 x 2	9 x 1	9 x 0	9 x 8	8 x 8	7 x 8	6 x 8
9 x 9	9 x 8	9 x 7	9 x 6	9 x 5	9 x 4	9 x 3

Day 16

Name:..........................

Score: /20

Time: ....

0 x9	1 x9	2 x9	3 x9	4 x9	5 x9	6 x9
7 x9	8 x9	9 x9	1 x2	2 x3	3 x4	4 x5
5 x6	6 x7	7 x8	8 x9	9 x1	2 x8	3 x7
4 x6	9 x0	0 x2	2 x4	8 x1	8 x6	5 x2
4 x0	9 x1	5 x3	1 x3	1 x6	7 x2	5 x1
4 x1	4 x2	8 x3	9 x3	4 x0	4 x3	4 x4
4 x8	4 x9	4 x5	4 x6	4 x7	0 x4	1 x4
5 x4	6 x4	2 x4	3 x4	4 x4	7 x4	8 x4
2 x5	3 x5	9 x4	5 x0	1 x5	4 x5	5 x5

Name:..........................

Score: /20

Time:
....

0	0	0	0	0	0	0
x 6	x 5	x 4	x 3	x 2	x 1	x 0

3	2	1	0	0	0	0
x 0	x 0	x 0	x 0	x 9	x 8	x 7

1	9	8	7	6	5	4
x 1	x 0	x 0	x 0	x 0	x 0	x 0

1	1	1	1	1	1	1
x 8	x 7	x 6	x 5	x 4	x 3	x 2

5	4	3	2	1	0	1
x 1	x 1	x 1	x 1	x 1	x 1	x 9

2	2	2	9	8	7	6
x 2	x 1	x 0	x 1	x 1	x 1	x 1

2	2	2	2	2	2	2
x 9	x 8	x 7	x 6	x 5	x 4	x 3

6	5	4	3	2	1	0
x 2	x 2	x 2	x 2	x 2	x 2	x 2

3	3	3	3	9	8	7
x 3	x 2	x 1	x 0	x 2	x 2	x 2

Day 18

Name:..............................

Score: /20

Time: ....

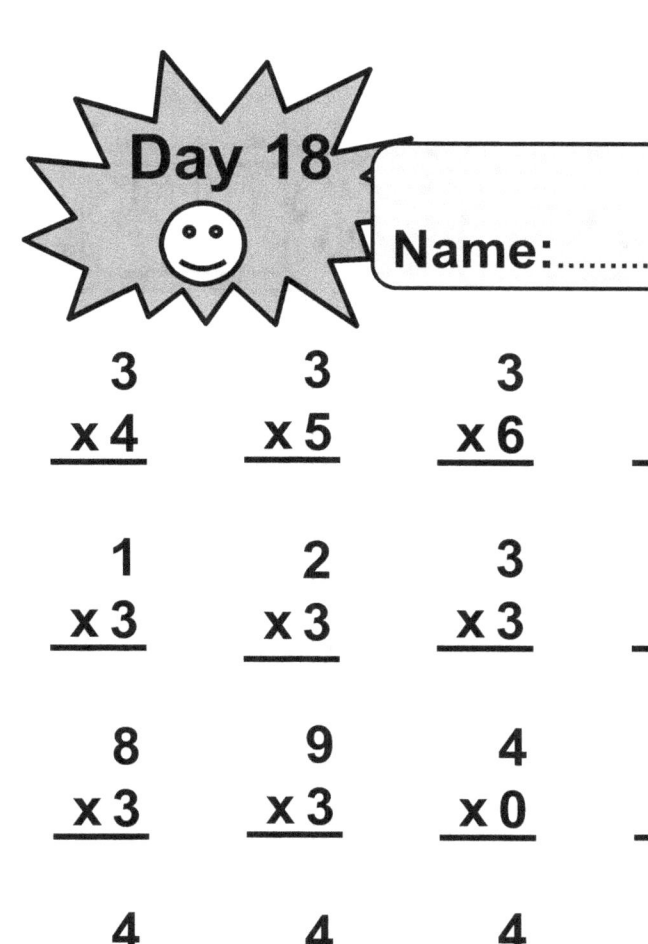

3 x 4	3 x 5	3 x 6	3 x 7	3 x 8	3 x 9	0 x 3
1 x 3	2 x 3	3 x 3	4 x 3	5 x 3	6 x 3	7 x 3
8 x 3	9 x 3	4 x 0	4 x 1	4 x 2	4 x 3	4 x 4
4 x 5	4 x 6	4 x 7	4 x 8	4 x 9	0 x 4	1 x 4
2 x 4	3 x 4	4 x 4	5 x 4	6 x 4	7 x 4	8 x 4
9 x 4	5 x 0	1 x 5	2 x 5	3 x 5	4 x 5	5 x 5
6 x 5	7 x 5	8 x 5	9 x 5	5 x 0	5 x 1	5 x 2
5 x 3	5 x 4	5 x 5	5 x 6	5 x 7	5 x 8	5 x 9
6 x 0	6 x 1	6 x 2	6 x 3	6 x 4	6 x 5	6 x 6

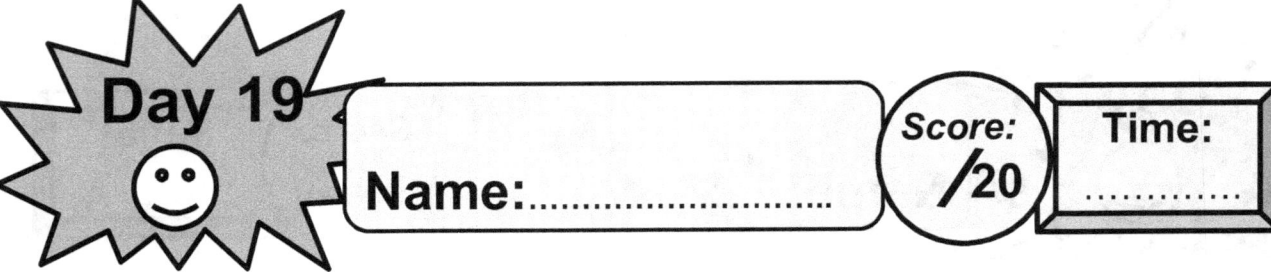

3 x 6	2 x 6	1 x 6	0 x 6	6 x 9	6 x 8	6 x 7
7 x 0	9 x 6	8 x 6	7 x 6	6 x 6	5 x 6	4 x 6
7 x 7	7 x 6	7 x 5	7 x 4	7 x 3	7 x 2	7 x 1
4 x 7	3 x 7	2 x 7	1 x 7	0 x 7	7 x 9	7 x 8
8 x 1	8 x 0	9 x 7	8 x 7	7 x 7	6 x 7	5 x 7
8 x 8	8 x 7	8 x 6	8 x 5	8 x 4	8 x 3	8 x 2
5 x 8	4 x 8	3 x 8	2 x 8	1 x 8	0 x 8	8 x 9
9 x 2	9 x 1	9 x 0	9 x 8	8 x 8	7 x 8	6 x 8
9 x 9	9 x 8	9 x 7	9 x 6	9 x 5	9 x 4	9 x 3

Day 20

Name:............................

Score: /20

Time: ....

0 x9	1 x9	2 x9	3 x9	4 x9	5 x9	6 x9
7 x9	8 x9	9 x9	1 x2	2 x3	3 x4	4 x5
5 x6	6 x7	7 x8	8 x9	9 x1	2 x8	3 x7
4 x6	9 x0	0 x2	2 x4	8 x1	8 x6	5 x2
4 x0	9 x1	5 x3	1 x3	1 x6	7 x2	5 x1
4 x1	4 x2	8 x3	9 x3	4 x0	4 x3	4 x4
4 x8	4 x9	4 x5	4 x6	4 x7	0 x4	1 x4
5 x4	6 x4	2 x4	3 x4	4 x4	7 x4	8 x4
2 x5	3 x5	9 x4	5 x0	1 x5	4 x5	5 x5

Name:.........................

Score: /20

Time: ....

1 x1	9 x0	8 x0	0 x3	0 x2	0 x1	0 x0
1 x8	1 x7	1 x6	0 x0	0 x9	0 x8	0 x7
5 x1	4 x1	3 x1	7 x0	6 x0	5 x0	4 x0
2 x2	2 x1	2 x0	1 x5	1 x4	1 x3	1 x2
2 x1	1 x1	0 x1	1 x9	0 x6	0 x5	0 x4
9 x1	8 x1	7 x1	6 x1	3 x0	2 x0	1 x0
2 x9	2 x8	2 x7	2 x6	2 x5	2 x4	2 x3
6 x2	5 x2	4 x2	3 x2	2 x2	1 x2	0 x2
3 x3	3 x2	3 x1	3 x0	9 x2	8 x2	7 x2

Name:........................

Score: /20

Time: ....

2	3	1	4	3	3	0
x 4	x 4	x 6	x 7	x 8	x 0	x 3

1	2	3	4	5	6	7
x 1	x 3	x 3	x 3	x 3	x 3	x 3

8	9	4	4	2	4	4
x 3	x 9	x 0	x 1	x 2	x 3	x 4

4	9	4	4	7	0	1
x 5	x 6	x 7	x 0	x 9	x 4	x 4

2	3	4	5	6	7	8
x 4	x 4	x 4	x 1	x 6	x 4	x 0

7						
9	5	1	2	3	4	5
x 4	x 0	x 5	x 7	x 5	x 4	x 5

6	7	8	9	5	6	5
x 5	x 9	x 5	x 7	x 0	x 1	x 2

2	5	5	1	5	5	5
x 3	x 4	x 5	x 6	x 7	x 8	x 9

6	5	6	7	6	8	6
x 0	x 1	x 2	x 3	x 4	x 5	x 6

Day 23

Name:...........................

Score: /20

Time: ....

3 x 0	2 x 6	1 x 9	0 x 6	6 x 8	6 x 7	6 x 7
7 x 0	9 x 1	8 x 6	7 x 2	6 x 6	5 x 3	4 x 6
5 x 2	4 x 6	6 x 5	7 x 4	8 x 3	7 x 2	9 x 1
4 x 0	3 x 7	2 x 9	1 x 7	0 x 2	7 x 9	7 x 8
8 x 1	8 x 0	9 x 7	8 x 7	7 x 7	6 x 7	5 x 7
8 x 8	5 x 7	8 x 6	2 x 5	8 x 4	9 x 3	8 x 2
6 x 8	4 x 5	3 x 8	2 x 4	1 x 8	0 x 8	8 x 9
9 x 2	9 x 1	9 x 0	9 x 8	8 x 8	7 x 8	6 x 8
9 x 9	9 x 8	9 x 7	9 x 6	9 x 5	9 x 4	9 x 3

Day 24

Name:..........................

Score: /20

Time: ....

0 x 9	1 x 9	2 x 9	3 x 9	4 x 9	5 x 9	6 x 9
7 x 9	8 x 9	9 x 9	1 x 2	2 x 3	3 x 4	4 x 5
5 x 6	6 x 7	7 x 8	8 x 9	9 x 1	2 x 8	3 x 7
4 x 6	9 x 0	0 x 2	2 x 4	8 x 1	8 x 6	5 x 2
4 x 0	9 x 1	5 x 3	1 x 3	1 x 6	7 x 2	5 x 1
4 x 1	4 x 2	8 x 3	9 x 3	4 x 0	4 x 3	4 x 4
4 x 8	4 x 9	4 x 5	4 x 6	4 x 7	0 x 4	1 x 4
5 x 4	6 x 4	2 x 4	3 x 4	4 x 4	7 x 4	8 x 4
2 x 5	3 x 5	9 x 4	5 x 0	1 x 5	4 x 5	5 x 5

Name:................................

Score: /20

Time: ....

0 x 6	0 x 5	0 x 4	0 x 3	0 x 2	0 x 1	0 x 0
3 x 0	2 x 0	1 x 0	0 x 0	0 x 9	0 x 8	0 x 7
1 x 1	9 x 0	8 x 0	7 x 0	6 x 0	5 x 0	4 x 0
1 x 8	1 x 7	1 x 6	1 x 5	1 x 4	1 x 3	1 x 2
5 x 1	4 x 1	3 x 1	2 x 1	1 x 1	0 x 1	1 x 9
2 x 2	2 x 1	2 x 0	9 x 1	8 x 1	7 x 1	6 x 1
2 x 9	2 x 8	2 x 7	2 x 6	2 x 5	2 x 4	2 x 3
6 x 2	5 x 2	4 x 2	3 x 2	2 x 2	1 x 2	0 x 2
3 x 3	3 x 2	3 x 1	3 x 0	9 x 2	8 x 2	7 x 2

Name:..............................

Score: /20

Time: ....

3 x 4	3 x 5	3 x 6	3 x 7	3 x 8	3 x 9	0 x 3
1 x 3	2 x 3	3 x 3	4 x 3	5 x 3	6 x 3	7 x 3
8 x 3	9 x 3	4 x 0	4 x 1	4 x 2	4 x 3	4 x 4
4 x 5	4 x 6	4 x 7	4 x 8	4 x 9	0 x 4	1 x 4
2 x 4	3 x 4	4 x 4	5 x 4	6 x 4	7 x 4	8 x 4
9 x 4	5 x 0	1 x 5	2 x 5	3 x 5	4 x 5	5 x 5
6 x 5	7 x 5	8 x 5	9 x 5	5 x 0	5 x 1	5 x 2
5 x 3	5 x 4	5 x 5	5 x 6	5 x 7	5 x 8	5 x 9
6 x 0	6 x 1	6 x 2	6 x 3	6 x 4	6 x 5	6 x 6

Name:...........................

Score: /20

Time: ....

3 x 6	2 x 6	1 x 6	0 x 6	6 x 9	6 x 8	6 x 7
7 x 0	9 x 6	8 x 6	7 x 6	6 x 6	5 x 6	4 x 6
7 x 7	7 x 6	7 x 5	7 x 4	7 x 3	7 x 2	7 x 1
4 x 7	3 x 7	2 x 7	1 x 7	0 x 7	7 x 9	7 x 8
8 x 1	8 x 0	9 x 7	8 x 7	7 x 7	6 x 7	5 x 7
8 x 8	8 x 7	8 x 6	8 x 5	8 x 4	8 x 3	8 x 2
5 x 8	4 x 8	3 x 8	2 x 8	1 x 8	0 x 8	8 x 9
9 x 2	9 x 1	9 x 0	9 x 8	8 x 8	7 x 8	6 x 8
9 x 9	9 x 8	9 x 7	9 x 6	9 x 5	9 x 4	9 x 3

Day 28

Name:....................

Score: /20

Time: ....

0	1	2	3	4	5	6
x9	x9	x9	x9	x9	x9	x9

7	8	9	1	2	3	4
x9	x9	x9	x2	x3	x4	x5

5	6	7	8	9	2	3
x6	x7	x8	x9	x1	x8	x7

4	9	0	2	8	8	5
x6	x0	x2	x4	x1	x6	x2

4	9	5	1	1	7	5
x0	x1	x3	x3	x6	x2	x1

4	4	8	9	4	4	4
x1	x2	x3	x3	x0	x3	x4

4	4	4	4	4	0	1
x8	x9	x5	x6	x7	x4	x4

5	6	2	3	4	7	8
x4	x4	x4	x4	x4	x4	x4

2	3	9	5	1	4	5
x5	x5	x4	x0	x5	x5	x5

Day 29

Name:..........................

Score: /20

Time: ....

1 x 1	9 x 0	8 x 0	0 x 3	0 x 2	0 x 1	0 x 0
1 x 8	1 x 7	1 x 6	0 x 0	0 x 9	0 x 8	0 x 7
5 x 1	4 x 1	3 x 1	7 x 0	6 x 0	5 x 0	4 x 0
2 x 2	2 x 1	2 x 0	1 x 5	1 x 4	1 x 3	1 x 2
2 x 1	1 x 1	0 x 1	1 x 9	0 x 6	0 x 5	0 x 4
9 x 1	8 x 1	7 x 1	6 x 1	3 x 0	2 x 0	1 x 0
2 x 9	2 x 8	2 x 7	2 x 6	2 x 5	2 x 4	2 x 3
6 x 2	5 x 2	4 x 2	3 x 2	2 x 2	1 x 2	0 x 2
3 x 3	3 x 2	3 x 1	3 x 0	9 x 2	8 x 2	7 x 2

Day 30

Name:.............................

Score: /20

Time: ....

2	3	1	4	3	3	0
x 4	x 4	x 6	x 7	x 8	x 0	x 3

1	2	3	4	5	6	7
x 1	x 3	x 3	x 3	x 3	x 3	x 3

8	9	4	4	2	4	4
x 3	x 9	x 0	x 1	x 2	x 3	x 4

4	9	4	4	7	0	1
x 5	x 6	x 7	x 0	x 9	x 4	x 4

2	3	4	5	6	7	8
x 4	x 4	x 4	x 1	x 6	x 4	x 0

7						
9	5	1	2	3	4	5
x 4	x 0	x 5	x 7	x 5	x 4	x 5

6	7	8	9	5	6	5
x 5	x 9	x 5	x 7	x 0	x 1	x 2

2	5	5	1	5	5	5
x 3	x 4	x 5	x 6	x 7	x 8	x 9

6	5	6	7	6	8	6
x 0	x 1	x 2	x 3	x 4	x 5	x 6

3	2	1	0	6	6	6
x 0	x 6	x 9	x 6	x 8	x 7	x 7

7	9	8	7	6	5	4
x 0	x 1	x 6	x 2	x 6	x 3	x 6

5	4	6	7	8	7	9
x 2	x 6	x 5	x 4	x 3	x 2	x 1

4	3	2	1	0	7	7
x 0	x 7	x 9	x 7	x 2	x 9	x 8

8	8	9	8	7	6	5
x 1	x 0	x 7	x 7	x 7	x 7	x 7

8	5	8	2	8	9	8
x 8	x 7	x 6	x 5	x 4	x 3	x 2

6	4	3	2	1	0	8
x 8	x 5	x 8	x 4	x 8	x 8	x 9

9	9	9	9	8	7	6
x 2	x 1	x 0	x 8	x 8	x 8	x 8

9	9	9	9	9	9	9
x 9	x 8	x 7	x 6	x 5	x 4	x 3

Day 32

Name:.............................

Score: /20

Time: ....

0	1	2	3	4	5	6
x9	x9	x9	x9	x9	x9	x9

7	8	9	1	2	3	4
x9	x9	x9	x2	x3	x4	x5

5	6	7	8	9	2	3
x6	x7	x8	x9	x1	x8	x7

4	9	0	2	8	8	5
x6	x0	x2	x4	x1	x6	x2

4	9	5	1	1	7	5
x0	x1	x3	x3	x6	x2	x1

4	4	8	9	4	4	4
x1	x2	x3	x3	x0	x3	x4

4	4	4	4	4	0	1
x8	x9	x5	x6	x7	x4	x4

5	6	2	3	4	7	8
x4	x4	x4	x4	x4	x4	x4

2	3	9	5	1	4	5
x5	x5	x4	x0	x5	x5	x5

Name:..........................

Score: /20

Time: ....

0 x 6	0 x 5	0 x 4	0 x 3	0 x 2	0 x 1	0 x 0
3 x 0	2 x 0	1 x 0	0 x 0	0 x 9	0 x 8	0 x 7
1 x 1	9 x 0	8 x 0	7 x 0	6 x 0	5 x 0	4 x 0
1 x 8	1 x 7	1 x 6	1 x 5	1 x 4	1 x 3	1 x 2
5 x 1	4 x 1	3 x 1	2 x 1	1 x 1	0 x 1	1 x 9
2 x 2	2 x 1	2 x 0	9 x 1	8 x 1	7 x 1	6 x 1
2 x 9	2 x 8	2 x 7	2 x 6	2 x 5	2 x 4	2 x 3
6 x 2	5 x 2	4 x 2	3 x 2	2 x 2	1 x 2	0 x 2
3 x 3	3 x 2	3 x 1	3 x 0	9 x 2	8 x 2	7 x 2

Name:...........................

Score: /20

Time: ....

3 x 4	3 x 5	3 x 6	3 x 7	3 x 8	3 x 9	0 x 3
1 x 3	2 x 3	3 x 3	4 x 3	5 x 3	6 x 3	7 x 3
8 x 3	9 x 3	4 x 0	4 x 1	4 x 2	4 x 3	4 x 4
4 x 5	4 x 6	4 x 7	4 x 8	4 x 9	0 x 4	1 x 4
2 x 4	3 x 4	4 x 4	5 x 4	6 x 4	7 x 4	8 x 4
9 x 4	5 x 0	1 x 5	2 x 5	3 x 5	4 x 5	5 x 5
6 x 5	7 x 5	8 x 5	9 x 5	5 x 0	5 x 1	5 x 2
5 x 3	5 x 4	5 x 5	5 x 6	5 x 7	5 x 8	5 x 9
6 x 0	6 x 1	6 x 2	6 x 3	6 x 4	6 x 5	6 x 6

Day 35

Name:.............................

Score: /20

Time: ....

3	2	1	0	6	6	6
x 6	x 6	x 6	x 6	x 9	x 8	x 7

7	9	8	7	6	5	4
x 0	x 6	x 6	x 6	x 6	x 6	x 6

7	7	7	7	7	7	7
x 7	x 6	x 5	x 4	x 3	x 2	x 1

4	3	2	1	0	7	7
x 7	x 7	x 7	x 7	x 7	x 9	x 8

8	8	9	8	7	6	5
x 1	x 0	x 7	x 7	x 7	x 7	x 7

8	8	8	8	8	8	8
x 8	x 7	x 6	x 5	x 4	x 3	x 2

5	4	3	2	1	0	8
x 8	x 8	x 8	x 8	x 8	x 8	x 9

9	9	9	9	8	7	6
x 2	x 1	x 0	x 8	x 8	x 8	x 8

9	9	9	9	9	9	9
x 9	x 8	x 7	x 6	x 5	x 4	x 3

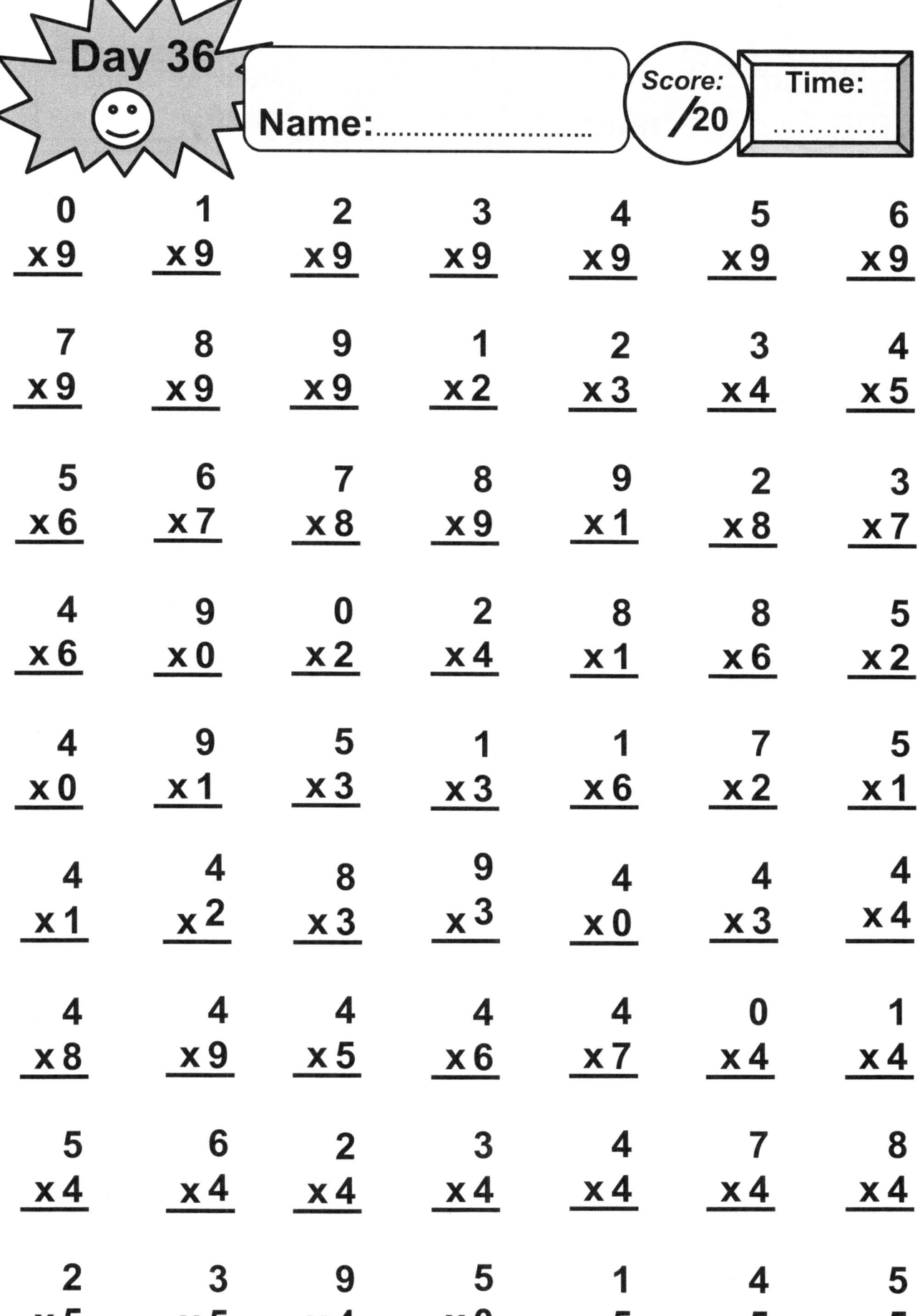

Day 36

Name:..................................

Score: /20

Time: ....

0	1	2	3	4	5	6
x9	x9	x9	x9	x9	x9	x9

7	8	9	1	2	3	4
x9	x9	x9	x2	x3	x4	x5

5	6	7	8	9	2	3
x6	x7	x8	x9	x1	x8	x7

4	9	0	2	8	8	5
x6	x0	x2	x4	x1	x6	x2

4	9	5	1	1	7	5
x0	x1	x3	x3	x6	x2	x1

4	4	8	9	4	4	4
x1	x2	x3	x3	x0	x3	x4

4	4	4	4	4	0	1
x8	x9	x5	x6	x7	x4	x4

5	6	2	3	4	7	8
x4	x4	x4	x4	x4	x4	x4

2	3	9	5	1	4	5
x5	x5	x4	x0	x5	x5	x5

Day 37

Name:...........................

Score: /20

Time: ....

1	9	8	0	0	0	0
x 1	x 0	x 0	x 3	x 2	x 1	x 0

1	1	1	0	0	0	0
x 8	x 7	x 6	x 0	x 9	x 8	x 7

5	4	3	7	6	5	4
x 1	x 1	x 1	x 0	x 0	x 0	x 0

2	2	2	1	1	1	1
x 2	x 1	x 0	x 5	x 4	x 3	x 2

2	1	0	1	0	0	0
x 1	x 1	x 1	x 9	x 6	x 5	x 4

9	8	7	6	3	2	1
x 1	x 1	x 1	x 1	x 0	x 0	x 0

2	2	2	2	2	2	2
x 9	x 8	x 7	x 6	x 5	x 4	x 3

6	5	4	3	2	1	0
x 2	x 2	x 2	x 2	x 2	x 2	x 2

3	3	3	3	9	8	7
x 3	x 2	x 1	x 0	x 2	x 2	x 2

Name:............................

Score: /20

Time: ....

2	3	1	4	3	3	0
x 4	x 4	x 6	x 7	x 8	x 0	x 3

1	2	3	4	5	6	7
x 1	x 3	x 3	x 3	x 3	x 3	x 3

8	9	4	4	2	4	4
x 3	x 9	x 0	x 1	x 2	x 3	x 4

4	9	4	4	7	0	1
x 5	x 6	x 7	x 0	x 9	x 4	x 4

2	3	4	5	6	7	8
x 4	x 4	x 4	x 1	x 6	x 4	x 0

7						
9	5	1	2	3	4	5
x 4	x 0	x 5	x 7	x 5	x 4	x 5

6	7	8	9	5	6	5
x 5	x 9	x 5	x 7	x 0	x 1	x 2

2	5	5	1	5	5	5
x 3	x 4	x 5	x 6	x 7	x 8	x 9

6	5	6	7	6	8	6
x 0	x 1	x 2	x 3	x 4	x 5	x 6

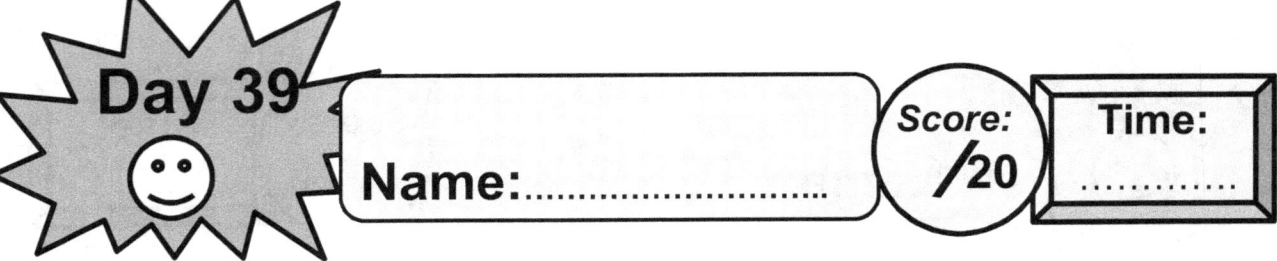

Name:...........................

Score: /20

Time: ....

3	2	1	0	6	6	6
x 0	x 6	x 9	x 6	x 8	x 7	x 7
7	9	8	7	6	5	4
x 0	x 1	x 6	x 2	x 6	x 3	x 6
5	4	6	7	8	7	9
x 2	x 6	x 5	x 4	x 3	x 2	x 1
4	3	2	1	0	7	7
x 0	x 7	x 9	x 7	x 2	x 9	x 8
8	8	9	8	7	6	5
x 1	x 0	x 7	x 7	x 7	x 7	x 7
8	5	8	2	8	9	8
x 8	x 7	x 6	x 5	x 4	x 3	x 2
6	4	3	2	1	0	8
x 8	x 5	x 8	x 4	x 8	x 8	x 9
9	9	9	9	8	7	6
x 2	x 1	x 0	x 8	x 8	x 8	x 8
9	9	9	9	9	9	9
x 9	x 8	x 7	x 6	x 5	x 4	x 3

Name:...........................

Score: /20

Time:
....

0	1	2	3	4	5	6
x9	x9	x9	x9	x9	x9	x9

7	8	9	1	2	3	4
x9	x9	x9	x2	x3	x4	x5

5	6	7	8	9	2	3
x6	x7	x8	x9	x1	x8	x7

4	9	0	2	8	8	5
x6	x0	x2	x4	x1	x6	x2

4	9	5	1	1	7	5
x0	x1	x3	x3	x6	x2	x1

4	4	8	9	4	4	4
x1	x2	x3	x3	x0	x3	x4

4	4	4	4	4	0	1
x8	x9	x5	x6	x7	x4	x4

5	6	2	3	4	7	8
x4	x4	x4	x4	x4	x4	x4

2	3	9	5	1	4	5
x5	x5	x4	x0	x5	x5	x5

Name:.............................

Score: /20

Time: ....

0 x 6	0 x 5	0 x 4	0 x 3	0 x 2	0 x 1	0 x 0
3 x 0	2 x 0	1 x 0	0 x 0	0 x 9	0 x 8	0 x 7
1 x 1	9 x 0	8 x 0	7 x 0	6 x 0	5 x 0	4 x 0
1 x 8	1 x 7	1 x 6	1 x 5	1 x 4	1 x 3	1 x 2
5 x 1	4 x 1	3 x 1	2 x 1	1 x 1	0 x 1	1 x 9
2 x 2	2 x 1	2 x 0	9 x 1	8 x 1	7 x 1	6 x 1
2 x 9	2 x 8	2 x 7	2 x 6	2 x 5	2 x 4	2 x 3
6 x 2	5 x 2	4 x 2	3 x 2	2 x 2	1 x 2	0 x 2
3 x 3	3 x 2	3 x 1	3 x 0	9 x 2	8 x 2	7 x 2

Day 42

Name:..........................

Score: /20

Time:
....

3	3	3	3	3	3	0
x 4	x 5	x 6	x 7	x 8	x 9	x 3

1	2	3	4	5	6	7
x 3	x 3	x 3	x 3	x 3	x 3	x 3

8	9	4	4	4	4	4
x 3	x 3	x 0	x 1	x 2	x 3	x 4

4	4	4	4	4	0	1
x 5	x 6	x 7	x 8	x 9	x 4	x 4

2	3	4	5	6	7	8
x 4	x 4	x 4	x 4	x 4	x 4	x 4

9	5	1	2	3	4	5
x 4	x 0	x 5	x 5	x 5	x 5	x 5

6	7	8	9	5	5	5
x 5	x 5	x 5	x 5	x 0	x 1	x 2

5	5	5	5	5	5	5
x 3	x 4	x 5	x 6	x 7	x 8	x 9

6	6	6	6	6	6	6
x 0	x 1	x 2	x 3	x 4	x 5	x 6

Name:................................

Score: /20

Time:
....

3 x 6	2 x 6	1 x 6	0 x 6	6 x 9	6 x 8	6 x 7
7 x 0	9 x 6	8 x 6	7 x 6	6 x 6	5 x 6	4 x 6
7 x 7	7 x 6	7 x 5	7 x 4	7 x 3	7 x 2	7 x 1
4 x 7	3 x 7	2 x 7	1 x 7	0 x 7	7 x 9	7 x 8
8 x 1	8 x 0	9 x 7	8 x 7	7 x 7	6 x 7	5 x 7
8 x 8	8 x 7	8 x 6	8 x 5	8 x 4	8 x 3	8 x 2
5 x 8	4 x 8	3 x 8	2 x 8	1 x 8	0 x 8	8 x 9
9 x 2	9 x 1	9 x 0	9 x 8	8 x 8	7 x 8	6 x 8
9 x 9	9 x 8	9 x 7	9 x 6	9 x 5	9 x 4	9 x 3

0	1	2	3	4	5	6
x 9	x 9	x 9	x 9	x 9	x 9	x 9

7	8	9	1	2	3	4
x 9	x 9	x 9	x 2	x 3	x 4	x 5

5	6	7	8	9	2	3
x 6	x 7	x 8	x 9	x 1	x 8	x 7

4	9	0	2	8	8	5
x 6	x 0	x 2	x 4	x 1	x 6	x 2

4	9	5	1	1	7	5
x 0	x 1	x 3	x 3	x 6	x 2	x 1

4	4	8	9	4	4	4
x 1	x 2	x 3	x 3	x 0	x 3	x 4

4	4	4	4	4	0	1
x 8	x 9	x 5	x 6	x 7	x 4	x 4

5	6	2	3	4	7	8
x 4	x 4	x 4	x 4	x 4	x 4	x 4

2	3	9	5	1	4	5
x 5	x 5	x 4	x 0	x 5	x 5	x 5

Day 45

1	9	8	0	0	0	0
x 1	x 0	x 0	x 3	x 2	x 1	x 0

1	1	1	0	0	0	0
x 8	x 7	x 6	x 0	x 9	x 8	x 7

5	4	3	7	6	5	4
x 1	x 1	x 1	x 0	x 0	x 0	x 0

2	2	2	1	1	1	1
x 2	x 1	x 0	x 5	x 4	x 3	x 2

2	1	0	1	0	0	0
x 1	x 1	x 1	x 9	x 6	x 5	x 4

9	8	7	6	3	2	1
x 1	x 1	x 1	x 1	x 0	x 0	x 0

2	2	2	2	2	2	2
x 9	x 8	x 7	x 6	x 5	x 4	x 3

6	5	4	3	2	1	0
x 2	x 2	x 2	x 2	x 2	x 2	x 2

3	3	3	3	9	8	7
x 3	x 2	x 1	x 0	x 2	x 2	x 2

Day 46

2	3	1	4	3	3	0
x 4	x 4	x 6	x 7	x 8	x 0	x 3

1	2	3	4	5	6	7
x 1	x 3	x 3	x 3	x 3	x 3	x 3

8	9	4	4	2	4	4
x 3	x 9	x 0	x 1	x 2	x 3	x 4

4	9	4	4	7	0	1
x 5	x 6	x 7	x 0	x 9	x 4	x 4

2	3	4	5	6	7	8
x 4	x 4	x 4	x 1	x 6	x 4	x 0

7

9	5	1	2	3	4	5
x 4	x 0	x 5	x 7	x 5	x 4	x 5

6	7	8	9	5	6	5
x 5	x 9	x 5	x 7	x 0	x 1	x 2

2	5	5	1	5	5	5
x 3	x 4	x 5	x 6	x 7	x 8	x 9

6	5	6	7	6	8	6
x 0	x 1	x 2	x 3	x 4	x 5	x 6

Day 47

Name:................................

Score: /20

Time: ....

3 x 0	2 x 6	1 x 9	0 x 6	6 x 8	6 x 7	6 x 7
7 x 0	9 x 1	8 x 6	7 x 2	6 x 6	5 x 3	4 x 6
5 x 2	4 x 6	6 x 5	7 x 4	8 x 3	7 x 2	9 x 1
4 x 0	3 x 7	2 x 9	1 x 7	0 x 2	7 x 9	7 x 8
8 x 1	8 x 0	9 x 7	8 x 7	7 x 7	6 x 7	5 x 7
8 x 8	5 x 7	8 x 6	2 x 5	8 x 4	9 x 3	8 x 2
6 x 8	4 x 5	3 x 8	2 x 4	1 x 8	0 x 8	8 x 9
9 x 2	9 x 1	9 x 0	9 x 8	8 x 8	7 x 8	6 x 8
9 x 9	9 x 8	9 x 7	9 x 6	9 x 5	9 x 4	9 x 3

Day 48

Name:..........................

Score: /20

Time: ....

0	1	2	3	4	5	6
x 9	x 9	x 9	x 9	x 9	x 9	x 9

7	8	9	1	2	3	4
x 9	x 9	x 9	x 2	x 3	x 4	x 5

5	6	7	8	9	2	3
x 6	x 7	x 8	x 9	x 1	x 8	x 7

4	9	0	2	8	8	5
x 6	x 0	x 2	x 4	x 1	x 6	x 2

4	9	5	1	1	7	5
x 0	x 1	x 3	x 3	x 6	x 2	x 1

4	4	8	9	4	4	4
x 1	x 2	x 3	x 3	x 0	x 3	x 4

4	4	4	4	4	0	1
x 8	x 9	x 5	x 6	x 7	x 4	x 4

5	6	2	3	4	7	8
x 4	x 4	x 4	x 4	x 4	x 4	x 4

2	3	9	5	1	4	5
x 5	x 5	x 4	x 0	x 5	x 5	x 5

Name:...........................

Score: /20

Time: ....

0 x 6	0 x 5	0 x 4	0 x 3	0 x 2	0 x 1	0 x 0
3 x 0	2 x 0	1 x 0	0 x 0	0 x 9	0 x 8	0 x 7
1 x 1	9 x 0	8 x 0	7 x 0	6 x 0	5 x 0	4 x 0
1 x 8	1 x 7	1 x 6	1 x 5	1 x 4	1 x 3	1 x 2
5 x 1	4 x 1	3 x 1	2 x 1	1 x 1	0 x 1	1 x 9
2 x 2	2 x 1	2 x 0	9 x 1	8 x 1	7 x 1	6 x 1
2 x 9	2 x 8	2 x 7	2 x 6	2 x 5	2 x 4	2 x 3
6 x 2	5 x 2	4 x 2	3 x 2	2 x 2	1 x 2	0 x 2
3 x 3	3 x 2	3 x 1	3 x 0	9 x 2	8 x 2	7 x 2

Score: /20

Time: ....

Name:.........................

3	3	3	3	3	3	0
x 4	x 5	x 6	x 7	x 8	x 9	x 3

1	2	3	4	5	6	7
x 3	x 3	x 3	x 3	x 3	x 3	x 3

8	9	4	4	4	4	4
x 3	x 3	x 0	x 1	x 2	x 3	x 4

4	4	4	4	4	0	1
x 5	x 6	x 7	x 8	x 9	x 4	x 4

2	3	4	5	6	7	8
x 4	x 4	x 4	x 4	x 4	x 4	x 4

9	5	1	2	3	4	5
x 4	x 0	x 5	x 5	x 5	x 5	x 5

6	7	8	9	5	5	5
x 5	x 5	x 5	x 5	x 0	x 1	x 2

5	5	5	5	5	5	5
x 3	x 4	x 5	x 6	x 7	x 8	x 9

6	6	6	6	6	6	6
x 0	x 1	x 2	x 3	x 4	x 5	x 6

Name:.............................

Score: /20

Time: ....

3	2	1	0	6	6	6
x 6	x 6	x 6	x 6	x 9	x 8	x 7

7	9	8	7	6	5	4
x 0	x 6	x 6	x 6	x 6	x 6	x 6

7	7	7	7	7	7	7
x 7	x 6	x 5	x 4	x 3	x 2	x 1

4	3	2	1	0	7	7
x 7	x 7	x 7	x 7	x 7	x 9	x 8

8	8	9	8	7	6	5
x 1	x 0	x 7	x 7	x 7	x 7	x 7

8	8	8	8	8	8	8
x 8	x 7	x 6	x 5	x 4	x 3	x 2

5	4	3	2	1	0	8
x 8	x 8	x 8	x 8	x 8	x 8	x 9

9	9	9	9	8	7	6
x 2	x 1	x 0	x 8	x 8	x 8	x 8

9	9	9	9	9	9	9
x 9	x 8	x 7	x 6	x 5	x 4	x 3

Day 52

Name:...........................

Score: /20

Time: ....

0	1	2	3	4	5	6
x 9	x 9	x 9	x 9	x 9	x 9	x 9

7	8	9	1	2	3	4
x 9	x 9	x 9	x 2	x 3	x 4	x 5

5	6	7	8	9	2	3
x 6	x 7	x 8	x 9	x 1	x 8	x 7

4	9	0	2	8	8	5
x 6	x 0	x 2	x 4	x 1	x 6	x 2

4	9	5	1	1	7	5
x 0	x 1	x 3	x 3	x 6	x 2	x 1

4	4	8	9	4	4	4
x 1	x 2	x 3	x 3	x 0	x 3	x 4

4	4	4	4	4	0	1
x 8	x 9	x 5	x 6	x 7	x 4	x 4

5	6	2	3	4	7	8
x 4	x 4	x 4	x 4	x 4	x 4	x 4

2	3	9	5	1	4	5
x 5	x 5	x 4	x 0	x 5	x 5	x 5

Name:................................

Score: /20

Time: ....

1	9	8	0	0	0	0
x 1	x 0	x 0	x 3	x 2	x 1	x 0

1	1	1	0	0	0	0
x 8	x 7	x 6	x 0	x 9	x 8	x 7

5	4	3	7	6	5	4
x 1	x 1	x 1	x 0	x 0	x 0	x 0

2	2	2	1	1	1	1
x 2	x 1	x 0	x 5	x 4	x 3	x 2

2	1	0	1	0	0	0
x 1	x 1	x 1	x 9	x 6	x 5	x 4

9	8	7	6	3	2	1
x 1	x 1	x 1	x 1	x 0	x 0	x 0

2	2	2	2	2	2	2
x 9	x 8	x 7	x 6	x 5	x 4	x 3

6	5	4	3	2	1	0
x 2	x 2	x 2	x 2	x 2	x 2	x 2

3	3	3	3	9	8	7
x 3	x 2	x 1	x 0	x 2	x 2	x 2

Day 55

Score: /20

Time: …………

Name:………………………

2 x 4	3 x 4	1 x 6	4 x 7	3 x 8	3 x 0	0 x 3
1 x 1	2 x 3	3 x 3	4 x 3	5 x 3	6 x 3	7 x 3
8 x 3	9 x 9	4 x 0	4 x 1	2 x 2	4 x 3	4 x 4
4 x 5	9 x 6	4 x 7	4 x 0	7 x 9	0 x 4	1 x 4
2 x 4 7	3 x 4	4 x 4	5 x 1	6 x 6	7 x 4	8 x 0
9 x 4	5 x 0	1 x 5	2 x 7	3 x 5	4 x 4	5 x 5
6 x 5	7 x 9	8 x 5	9 x 7	5 x 0	6 x 1	5 x 2
2 x 3	5 x 4	5 x 5	1 x 6	5 x 7	5 x 8	5 x 9
6 x 0	5 x 1	6 x 2	7 x 3	6 x 4	8 x 5	6 x 6

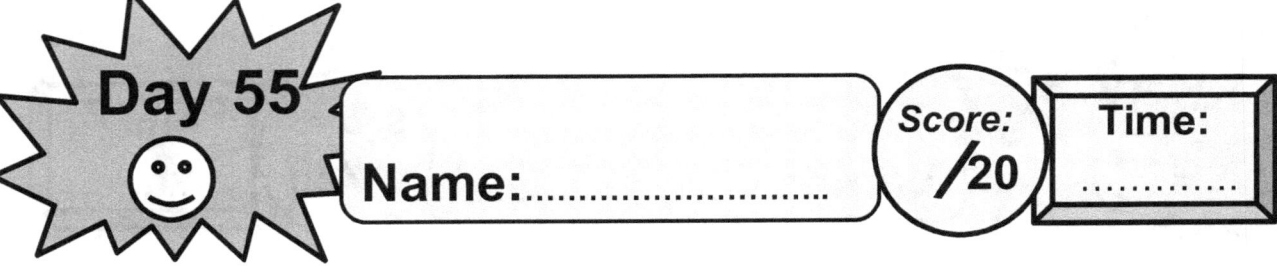
3 x 0	2 x 6	1 x 9	0 x 6	6 x 8	6 x 7	6 x 7
7 x 0	9 x 1	8 x 6	7 x 2	6 x 6	5 x 3	4 x 6
5 x 2	4 x 6	6 x 5	7 x 4	8 x 3	7 x 2	9 x 1
4 x 0	3 x 7	2 x 9	1 x 7	0 x 2	7 x 9	7 x 8
8 x 1	8 x 0	9 x 7	8 x 7	7 x 7	6 x 7	5 x 7
8 x 8	5 x 7	8 x 6	2 x 5	8 x 4	9 x 3	8 x 2
6 x 8	4 x 5	3 x 8	2 x 4	1 x 8	0 x 8	8 x 9
9 x 2	9 x 1	9 x 0	9 x 8	8 x 8	7 x 8	6 x 8
9 x 9	9 x 8	9 x 7	9 x 6	9 x 5	9 x 4	9 x 3

Day 56

Name:...............................

Score: /20

Time: ....

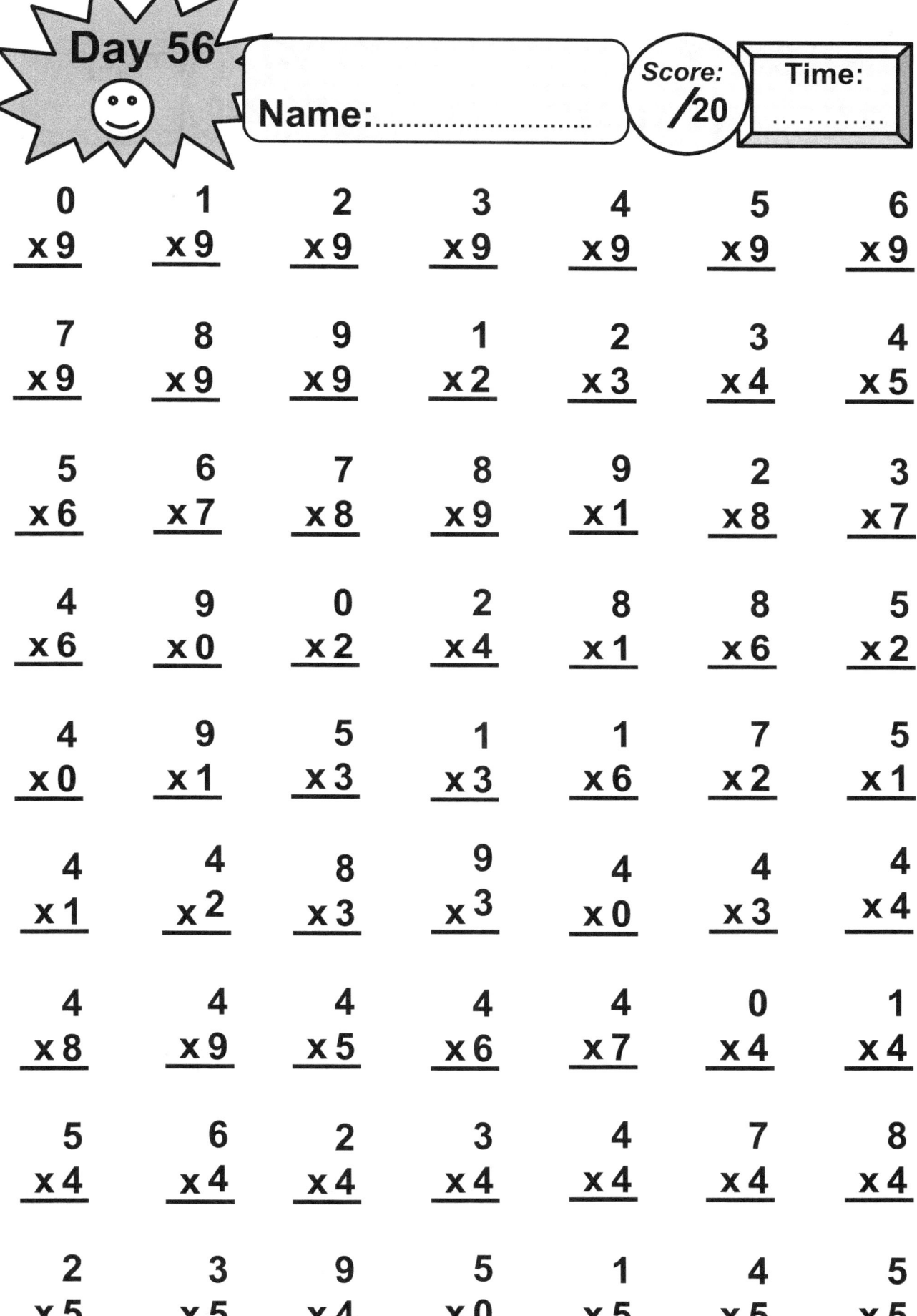

0 x 9	1 x 9	2 x 9	3 x 9	4 x 9	5 x 9	6 x 9
7 x 9	8 x 9	9 x 9	1 x 2	2 x 3	3 x 4	4 x 5
5 x 6	6 x 7	7 x 8	8 x 9	9 x 1	2 x 8	3 x 7
4 x 6	9 x 0	0 x 2	2 x 4	8 x 1	8 x 6	5 x 2
4 x 0	9 x 1	5 x 3	1 x 3	1 x 6	7 x 2	5 x 1
4 x 1	4 x 2	8 x 3	9 x 3	4 x 0	4 x 3	4 x 4
4 x 8	4 x 9	4 x 5	4 x 6	4 x 7	0 x 4	1 x 4
5 x 4	6 x 4	2 x 4	3 x 4	4 x 4	7 x 4	8 x 4
2 x 5	3 x 5	9 x 4	5 x 0	1 x 5	4 x 5	5 x 5

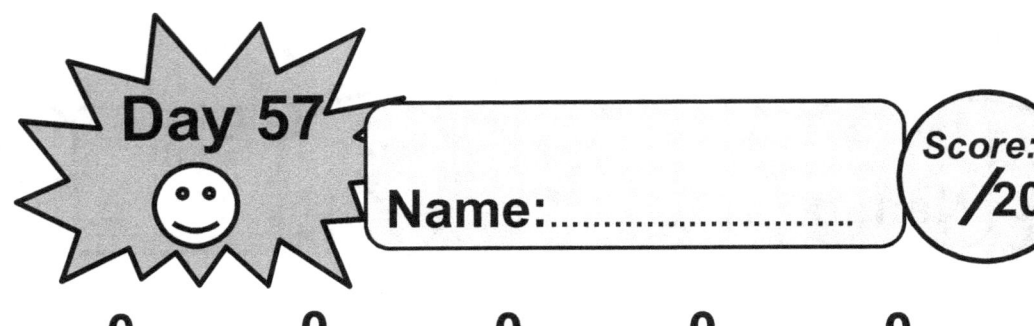

0	0	0	0	0	0	0
x 6	x 5	x 4	x 3	x 2	x 1	x 0

3	2	1	0	0	0	0
x 0	x 0	x 0	x 0	x 9	x 8	x 7

1	9	8	7	6	5	4
x 1	x 0	x 0	x 0	x 0	x 0	x 0

1	1	1	1	1	1	1
x 8	x 7	x 6	x 5	x 4	x 3	x 2

5	4	3	2	1	0	1
x 1	x 1	x 1	x 1	x 1	x 1	x 9

2	2	2	9	8	7	6
x 2	x 1	x 0	x 1	x 1	x 1	x 1

2	2	2	2	2	2	2
x 9	x 8	x 7	x 6	x 5	x 4	x 3

6	5	4	3	2	1	0
x 2	x 2	x 2	x 2	x 2	x 2	x 2

3	3	3	3	9	8	7
x 3	x 2	x 1	x 0	x 2	x 2	x 2

Day 58

3 x 4	3 x 5	3 x 6	3 x 7	3 x 8	3 x 9	0 x 3
1 x 3	2 x 3	3 x 3	4 x 3	5 x 3	6 x 3	7 x 3
8 x 3	9 x 3	4 x 0	4 x 1	4 x 2	4 x 3	4 x 4
4 x 5	4 x 6	4 x 7	4 x 8	4 x 9	0 x 4	1 x 4
2 x 4	3 x 4	4 x 4	5 x 4	6 x 4	7 x 4	8 x 4
9 x 4	5 x 0	1 x 5	2 x 5	3 x 5	4 x 5	5 x 5
6 x 5	7 x 5	8 x 5	9 x 5	5 x 0	5 x 1	5 x 2
5 x 3	5 x 4	5 x 5	5 x 6	5 x 7	5 x 8	5 x 9
6 x 0	6 x 1	6 x 2	6 x 3	6 x 4	6 x 5	6 x 6

Day 59

Name:................................

Score: /20

Time: ....

3 x 6	2 x 6	1 x 6	0 x 6	6 x 9	6 x 8	6 x 7
7 x 0	9 x 6	8 x 6	7 x 6	6 x 6	5 x 6	4 x 6
7 x 7	7 x 6	7 x 5	7 x 4	7 x 3	7 x 2	7 x 1
4 x 7	3 x 7	2 x 7	1 x 7	0 x 7	7 x 9	7 x 8
8 x 1	8 x 0	9 x 7	8 x 7	7 x 7	6 x 7	5 x 7
8 x 8	8 x 7	8 x 6	8 x 5	8 x 4	8 x 3	8 x 2
5 x 8	4 x 8	3 x 8	2 x 8	1 x 8	0 x 8	8 x 9
9 x 2	9 x 1	9 x 0	9 x 8	8 x 8	7 x 8	6 x 8
9 x 9	9 x 8	9 x 7	9 x 6	9 x 5	9 x 4	9 x 3

Day 60

Name:..........................

Score: /20

Time: ....

0	1	2	3	4	5	6
x 9	x 9	x 9	x 9	x 9	x 9	x 9

7	8	9	1	2	3	4
x 9	x 9	x 9	x 2	x 3	x 4	x 5

5	6	7	8	9	2	3
x 6	x 7	x 8	x 9	x 1	x 8	x 7

4	9	0	2	8	8	5
x 6	x 0	x 2	x 4	x 1	x 6	x 2

4	9	5	1	1	7	5
x 0	x 1	x 3	x 3	x 6	x 2	x 1

4	4	8	9	4	4	4
x 1	x 2	x 3	x 3	x 0	x 3	x 4

4	4	4	4	4	0	1
x 8	x 9	x 5	x 6	x 7	x 4	x 4

5	6	2	3	4	7	8
x 4	x 4	x 4	x 4	x 4	x 4	x 4

2	3	9	5	1	4	5
x 5	x 5	x 4	x 0	x 5	x 5	x 5

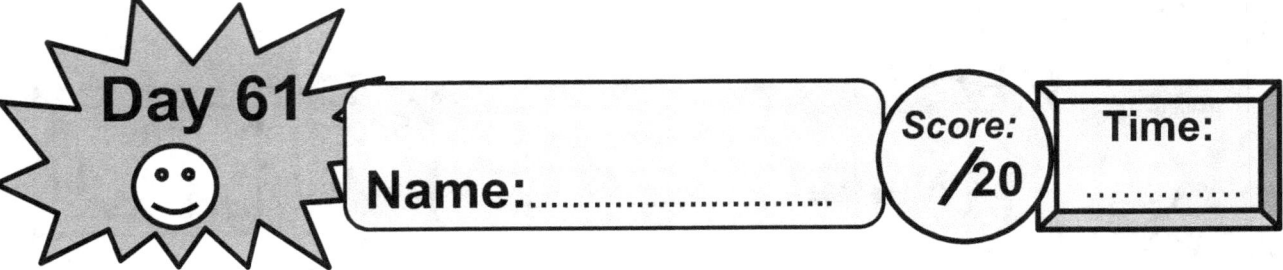

1 x 1	9 x 0	8 x 0	0 x 3	0 x 2	0 x 1	0 x 0
1 x 8	1 x 7	1 x 6	0 x 0	0 x 9	0 x 8	0 x 7
5 x 1	4 x 1	3 x 1	7 x 0	6 x 0	5 x 0	4 x 0
2 x 2	2 x 1	2 x 0	1 x 5	1 x 4	1 x 3	1 x 2
2 x 1	1 x 1	0 x 1	1 x 9	0 x 6	0 x 5	0 x 4
9 x 1	8 x 1	7 x 1	6 x 1	3 x 0	2 x 0	1 x 0
2 x 9	2 x 8	2 x 7	2 x 6	2 x 5	2 x 4	2 x 3
6 x 2	5 x 2	4 x 2	3 x 2	2 x 2	1 x 2	0 x 2
3 x 3	3 x 2	3 x 1	3 x 0	9 x 2	8 x 2	7 x 2

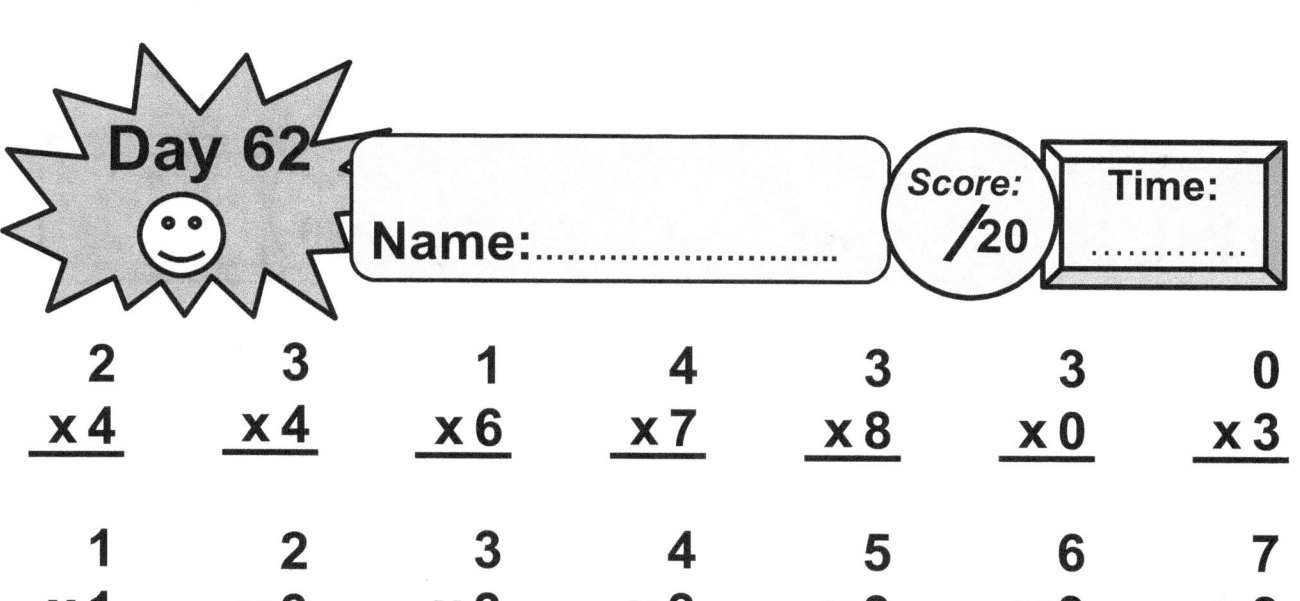

2 x 4	3 x 4	1 x 6	4 x 7	3 x 8	3 x 0	0 x 3
1 x 1	2 x 3	3 x 3	4 x 3	5 x 3	6 x 3	7 x 3
8 x 3	9 x 9	4 x 0	4 x 1	2 x 2	4 x 3	4 x 4
4 x 5	9 x 6	4 x 7	4 x 0	7 x 9	0 x 4	1 x 4
2 x 4	3 x 4	4 x 4	5 x 1	6 x 6	7 x 4	8 x 0
7 9 x 4	5 x 0	1 x 5	2 x 7	3 x 5	4 x 4	5 x 5
6 x 5	7 x 9	8 x 5	9 x 7	5 x 0	6 x 1	5 x 2
2 x 3	5 x 4	5 x 5	1 x 6	5 x 7	5 x 8	5 x 9
6 x 0	5 x 1	6 x 2	7 x 3	6 x 4	8 x 5	6 x 6

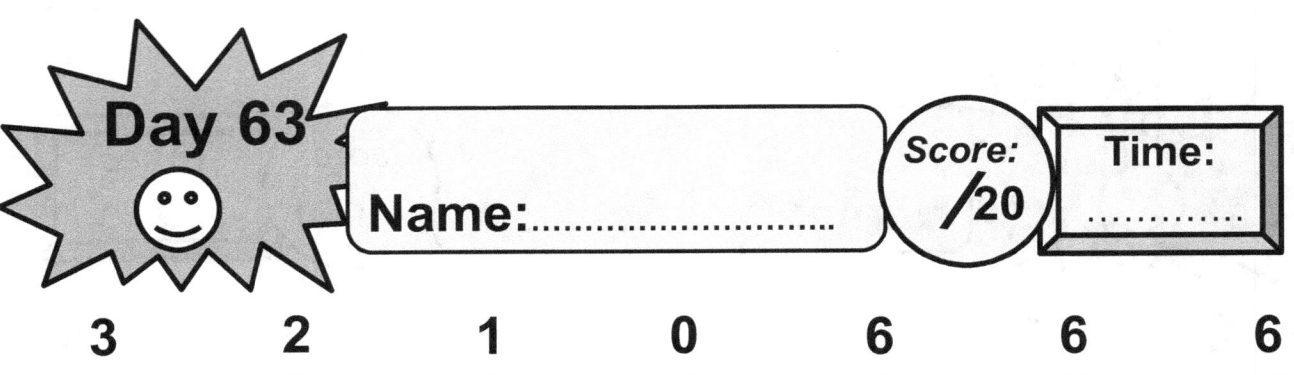

3 x 0	2 x 6	1 x 9	0 x 6	6 x 8	6 x 7	6 x 7
7 x 0	9 x 1	8 x 6	7 x 2	6 x 6	5 x 3	4 x 6
5 x 2	4 x 6	6 x 5	7 x 4	8 x 3	7 x 2	9 x 1
4 x 0	3 x 7	2 x 9	1 x 7	0 x 2	7 x 9	7 x 8
8 x 1	8 x 0	9 x 7	8 x 7	7 x 7	6 x 7	5 x 7
8 x 8	5 x 7	8 x 6	2 x 5	8 x 4	9 x 3	8 x 2
6 x 8	4 x 5	3 x 8	2 x 4	1 x 8	0 x 8	8 x 9
9 x 2	9 x 1	9 x 0	9 x 8	8 x 8	7 x 8	6 x 8
9 x 9	9 x 8	9 x 7	9 x 6	9 x 5	9 x 4	9 x 3

Name: ....

Score: /20

Time: ....

0	1	2	3	4	5	6
x9	x9	x9	x9	x9	x9	x9

7	8	9	1	2	3	4
x9	x9	x9	x2	x3	x4	x5

5	6	7	8	9	2	3
x6	x7	x8	x9	x1	x8	x7

4	9	0	2	8	8	5
x6	x0	x2	x4	x1	x6	x2

4	9	5	1	1	7	5
x0	x1	x3	x3	x6	x2	x1

4	4	8	9	4	4	4
x1	x2	x3	x3	x0	x3	x4

4	4	4	4	4	0	1
x8	x9	x5	x6	x7	x4	x4

5	6	2	3	4	7	8
x4	x4	x4	x4	x4	x4	x4

2	3	9	5	1	4	5
x5	x5	x4	x0	x5	x5	x5

Day 65

Name:........................

Score: /20

Time: ....

0	0	0	0	0	0	0
x 6	x 5	x 4	x 3	x 2	x 1	x 0

3	2	1	0	0	0	0
x 0	x 0	x 0	x 0	x 9	x 8	x 7

1	9	8	7	6	5	4
x 1	x 0	x 0	x 0	x 0	x 0	x 0

1	1	1	1	1	1	1
x 8	x 7	x 6	x 5	x 4	x 3	x 2

5	4	3	2	1	0	1
x 1	x 1	x 1	x 1	x 1	x 1	x 9

2	2	2	9	8	7	6
x 2	x 1	x 0	x 1	x 1	x 1	x 1

2	2	2	2	2	2	2
x 9	x 8	x 7	x 6	x 5	x 4	x 3

6	5	4	3	2	1	0
x 2	x 2	x 2	x 2	x 2	x 2	x 2

3	3	3	3	9	8	7
x 3	x 2	x 1	x 0	x 2	x 2	x 2

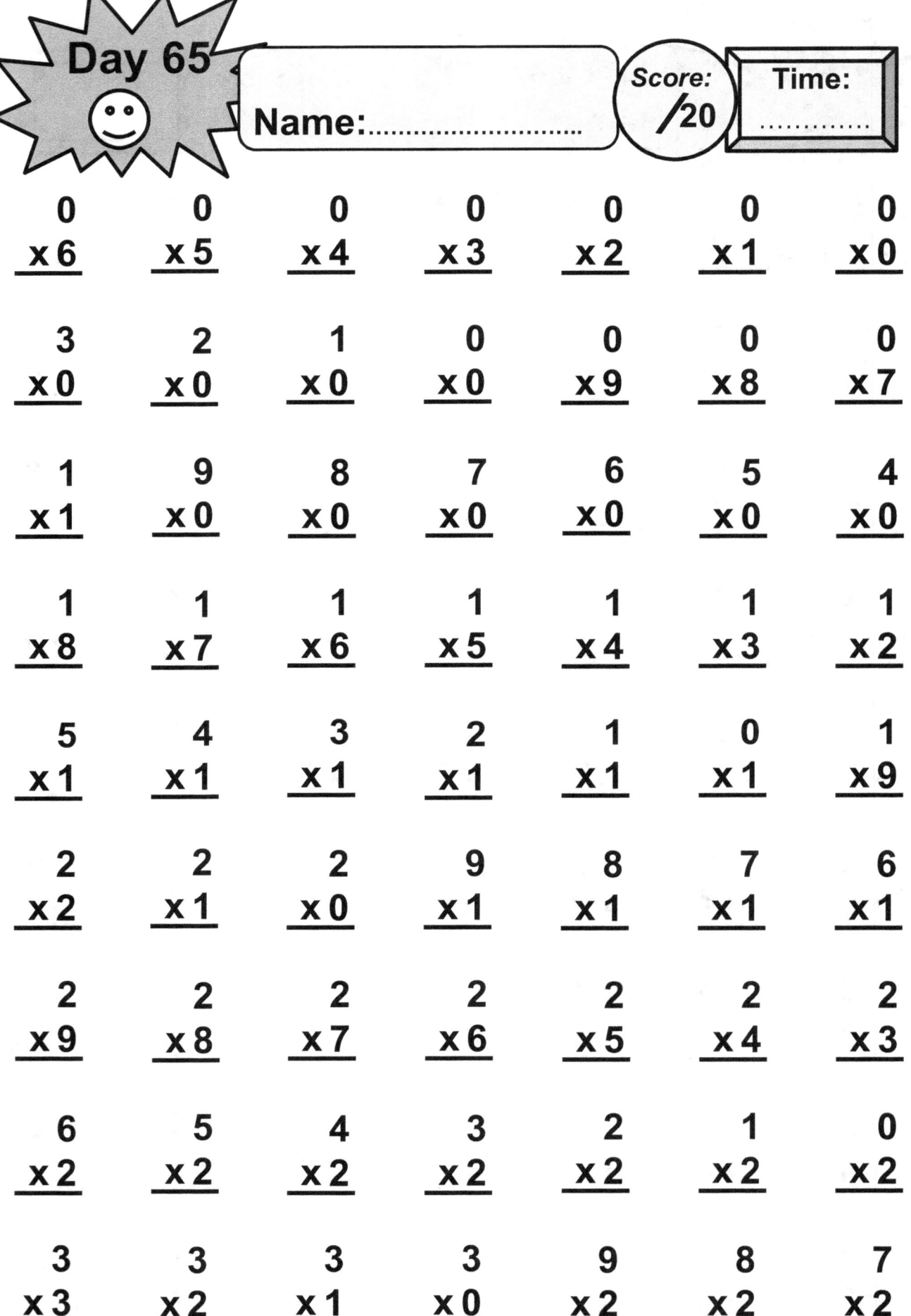

Name:..........................

Score: /20

Time: ....

3 x 4	3 x 5	3 x 6	3 x 7	3 x 8	3 x 9	0 x 3
1 x 3	2 x 3	3 x 3	4 x 3	5 x 3	6 x 3	7 x 3
8 x 3	9 x 3	4 x 0	4 x 1	4 x 2	4 x 3	4 x 4
4 x 5	4 x 6	4 x 7	4 x 8	4 x 9	0 x 4	1 x 4
2 x 4	3 x 4	4 x 4	5 x 4	6 x 4	7 x 4	8 x 4
9 x 4	5 x 0	1 x 5	2 x 5	3 x 5	4 x 5	5 x 5
6 x 5	7 x 5	8 x 5	9 x 5	5 x 0	5 x 1	5 x 2
5 x 3	5 x 4	5 x 5	5 x 6	5 x 7	5 x 8	5 x 9
6 x 0	6 x 1	6 x 2	6 x 3	6 x 4	6 x 5	6 x 6

Name:...........................

Score: /20

Time: ....

3 x 6	2 x 6	1 x 6	0 x 6	6 x 9	6 x 8	6 x 7
7 x 0	9 x 6	8 x 6	7 x 6	6 x 6	5 x 6	4 x 6
7 x 7	7 x 6	7 x 5	7 x 4	7 x 3	7 x 2	7 x 1
4 x 7	3 x 7	2 x 7	1 x 7	0 x 7	7 x 9	7 x 8
8 x 1	8 x 0	9 x 7	8 x 7	7 x 7	6 x 7	5 x 7
8 x 8	8 x 7	8 x 6	8 x 5	8 x 4	8 x 3	8 x 2
5 x 8	4 x 8	3 x 8	2 x 8	1 x 8	0 x 8	8 x 9
9 x 2	9 x 1	9 x 0	9 x 8	8 x 8	7 x 8	6 x 8
9 x 9	9 x 8	9 x 7	9 x 6	9 x 5	9 x 4	9 x 3

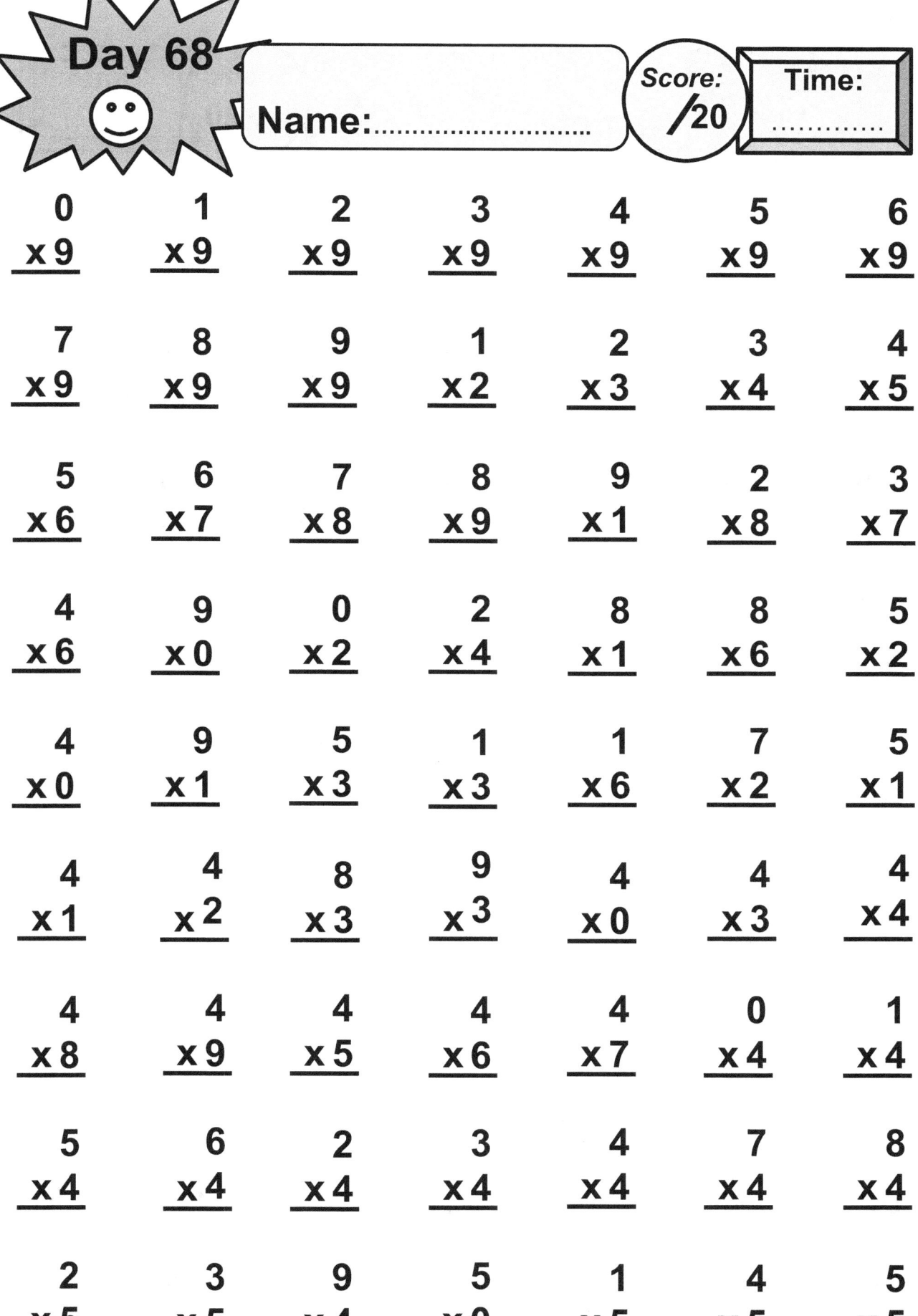

Day 68

Name:..............................

Score: /20

Time: ....

0	1	2	3	4	5	6
x9	x9	x9	x9	x9	x9	x9

7	8	9	1	2	3	4
x9	x9	x9	x2	x3	x4	x5

5	6	7	8	9	2	3
x6	x7	x8	x9	x1	x8	x7

4	9	0	2	8	8	5
x6	x0	x2	x4	x1	x6	x2

4	9	5	1	1	7	5
x0	x1	x3	x3	x6	x2	x1

4	4	8	9	4	4	4
x1	x2	x3	x3	x0	x3	x4

4	4	4	4	4	0	1
x8	x9	x5	x6	x7	x4	x4

5	6	2	3	4	7	8
x4	x4	x4	x4	x4	x4	x4

2	3	9	5	1	4	5
x5	x5	x4	x0	x5	x5	x5

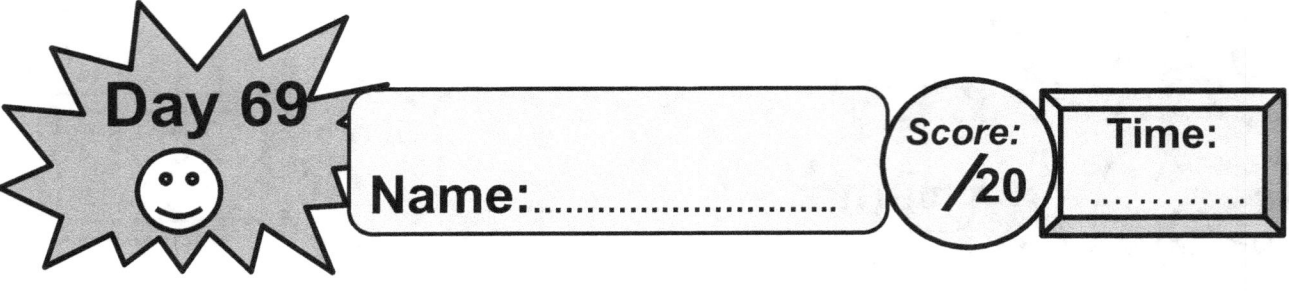

1	9	8	0	0	0	0
x 1	x 0	x 0	x 3	x 2	x 1	x 0

1	1	1	0	0	0	0
x 8	x 7	x 6	x 0	x 9	x 8	x 7

5	4	3	7	6	5	4
x 1	x 1	x 1	x 0	x 0	x 0	x 0

2	2	2	1	1	1	1
x 2	x 1	x 0	x 5	x 4	x 3	x 2

2	1	0	1	0	0	0
x 1	x 1	x 1	x 9	x 6	x 5	x 4

9	8	7	6	3	2	1
x 1	x 1	x 1	x 1	x 0	x 0	x 0

2	2	2	2	2	2	2
x 9	x 8	x 7	x 6	x 5	x 4	x 3

6	5	4	3	2	1	0
x 2	x 2	x 2	x 2	x 2	x 2	x 2

3	3	3	3	9	8	7
x 3	x 2	x 1	x 0	x 2	x 2	x 2

Day 70

Name:...........................

Score: /20

Time: ....

2	3	1	4	3	3	0
x 4	x 4	x 6	x 7	x 8	x 0	x 3

1	2	3	4	5	6	7
x 1	x 3	x 3	x 3	x 3	x 3	x 3

8	9	4	4	2	4	4
x 3	x 9	x 0	x 1	x 2	x 3	x 4

4	9	4	4	7	0	1
x 5	x 6	x 7	x 0	x 9	x 4	x 4

2	3	4	5	6	7	8
x 4	x 4	x 4	x 1	x 6	x 4	x 0
7						

9	5	1	2	3	4	5
x 4	x 0	x 5	x 7	x 5	x 4	x 5

6	7	8	9	5	6	5
x 5	x 9	x 5	x 7	x 0	x 1	x 2

2	5	5	1	5	5	5
x 3	x 4	x 5	x 6	x 7	x 8	x 9

6	5	6	7	6	8	6
x 0	x 1	x 2	x 3	x 4	x 5	x 6

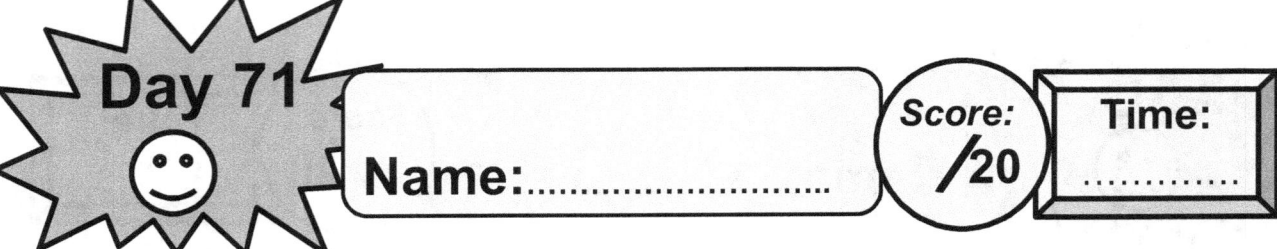

3 x 0	2 x 6	1 x 9	0 x 6	6 x 8	6 x 7	6 x 7
7 x 0	9 x 1	8 x 6	7 x 2	6 x 6	5 x 3	4 x 6
5 x 2	4 x 6	6 x 5	7 x 4	8 x 3	7 x 2	9 x 1
4 x 0	3 x 7	2 x 9	1 x 7	0 x 2	7 x 9	7 x 8
8 x 1	8 x 0	9 x 7	8 x 7	7 x 7	6 x 7	5 x 7
8 x 8	5 x 7	8 x 6	2 x 5	8 x 4	9 x 3	8 x 2
6 x 8	4 x 5	3 x 8	2 x 4	1 x 8	0 x 8	8 x 9
9 x 2	9 x 1	9 x 0	9 x 8	8 x 8	7 x 8	6 x 8
9 x 9	9 x 8	9 x 7	9 x 6	9 x 5	9 x 4	9 x 3

Name:...........................

Score: /20

Time: ....

0 x 9	1 x 9	2 x 9	3 x 9	4 x 9	5 x 9	6 x 9
7 x 9	8 x 9	9 x 9	1 x 2	2 x 3	3 x 4	4 x 5
5 x 6	6 x 7	7 x 8	8 x 9	9 x 1	2 x 8	3 x 7
4 x 6	9 x 0	0 x 2	2 x 4	8 x 1	8 x 6	5 x 2
4 x 0	9 x 1	5 x 3	1 x 3	1 x 6	7 x 2	5 x 1
4 x 1	4 x 2	8 x 3	9 x 3	4 x 0	4 x 3	4 x 4
4 x 8	4 x 9	4 x 5	4 x 6	4 x 7	0 x 4	1 x 4
5 x 4	6 x 4	2 x 4	3 x 4	4 x 4	7 x 4	8 x 4
2 x 5	3 x 5	9 x 4	5 x 0	1 x 5	4 x 5	5 x 5

Name:...........................

Score: /20

Time: ....

0	0	0	0	0	0	0
x 6	x 5	x 4	x 3	x 2	x 1	x 0

3	2	1	0	0	0	0
x 0	x 0	x 0	x 0	x 9	x 8	x 7

1	9	8	7	6	5	4
x 1	x 0	x 0	x 0	x 0	x 0	x 0

1	1	1	1	1	1	1
x 8	x 7	x 6	x 5	x 4	x 3	x 2

5	4	3	2	1	0	1
x 1	x 1	x 1	x 1	x 1	x 1	x 9

2	2	2	9	8	7	6
x 2	x 1	x 0	x 1	x 1	x 1	x 1

2	2	2	2	2	2	2
x 9	x 8	x 7	x 6	x 5	x 4	x 3

6	5	4	3	2	1	0
x 2	x 2	x 2	x 2	x 2	x 2	x 2

3	3	3	3	9	8	7
x 3	x 2	x 1	x 0	x 2	x 2	x 2

Name:..........................

Score: /20

Time:
....

3	3	3	3	3	3	0
x 4	x 5	x 6	x 7	x 8	x 9	x 3

1	2	3	4	5	6	7
x 3	x 3	x 3	x 3	x 3	x 3	x 3

8	9	4	4	4	4	4
x 3	x 3	x 0	x 1	x 2	x 3	x 4

4	4	4	4	4	0	1
x 5	x 6	x 7	x 8	x 9	x 4	x 4

2	3	4	5	6	7	8
x 4	x 4	x 4	x 4	x 4	x 4	x 4

9	5	1	2	3	4	5
x 4	x 0	x 5	x 5	x 5	x 5	x 5

6	7	8	9	5	5	5
x 5	x 5	x 5	x 5	x 0	x 1	x 2

5	5	5	5	5	5	5
x 3	x 4	x 5	x 6	x 7	x 8	x 9

6	6	6	6	6	6	6
x 0	x 1	x 2	x 3	x 4	x 5	x 6

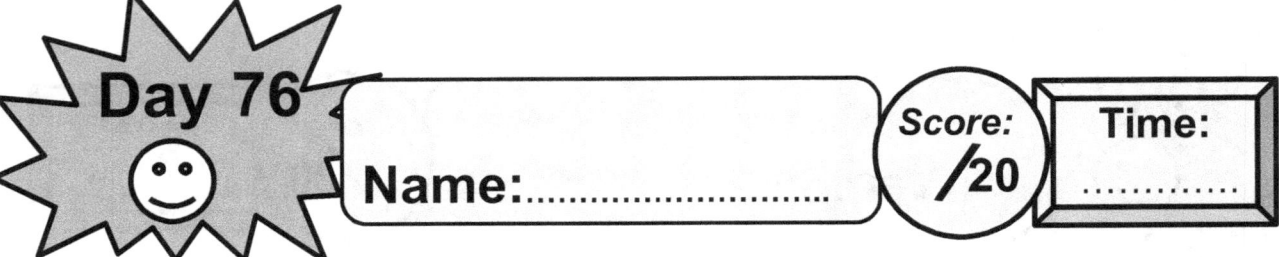

3 x6	2 x6	1 x6	0 x6	6 x9	6 x8	6 x7
7 x0	9 x6	8 x6	7 x6	6 x6	5 x6	4 x6
7 x7	7 x6	7 x5	7 x4	7 x3	7 x2	7 x1
4 x7	3 x7	2 x7	1 x7	0 x7	7 x9	7 x8
8 x1	8 x0	9 x7	8 x7	7 x7	6 x7	5 x7
8 x8	8 x7	8 x6	8 x5	8 x4	8 x3	8 x2
5 x8	4 x8	3 x8	2 x8	1 x8	0 x8	8 x9
9 x2	9 x1	9 x0	9 x8	8 x8	7 x8	6 x8
9 x9	9 x8	9 x7	9 x6	9 x5	9 x4	9 x3

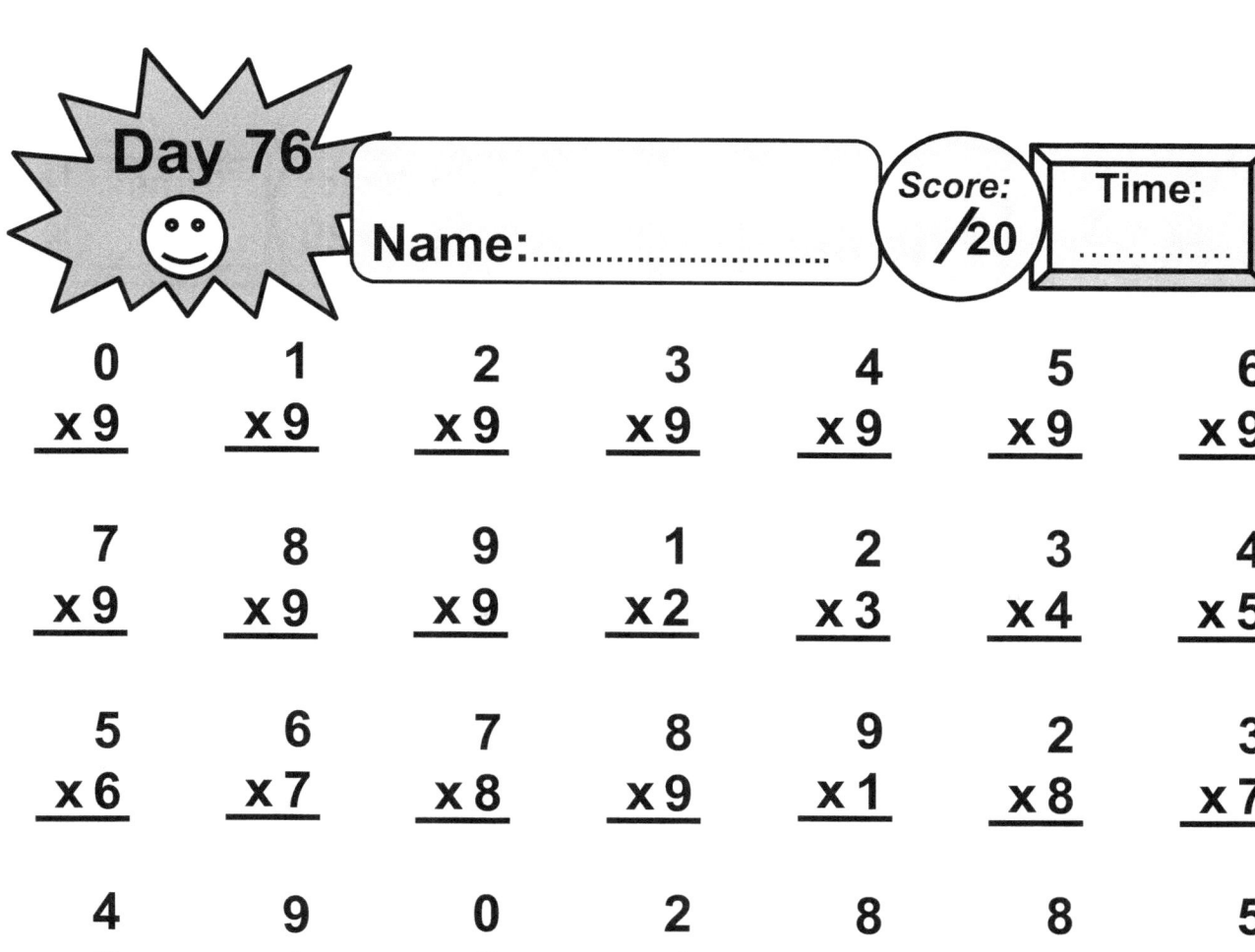

Day 76

Name:.............................

Score: /20

Time: ....

0	1	2	3	4	5	6
x 9	x 9	x 9	x 9	x 9	x 9	x 9

7	8	9	1	2	3	4
x 9	x 9	x 9	x 2	x 3	x 4	x 5

5	6	7	8	9	2	3
x 6	x 7	x 8	x 9	x 1	x 8	x 7

4	9	0	2	8	8	5
x 6	x 0	x 2	x 4	x 1	x 6	x 2

4	9	5	1	1	7	5
x 0	x 1	x 3	x 3	x 6	x 2	x 1

4	4	8	9	4	4	4
x 1	x 2	x 3	x 3	x 0	x 3	x 4

4	4	4	4	4	0	1
x 8	x 9	x 5	x 6	x 7	x 4	x 4

5	6	2	3	4	7	8
x 4	x 4	x 4	x 4	x 4	x 4	x 4

2	3	9	5	1	4	5
x 5	x 5	x 4	x 0	x 5	x 5	x 5

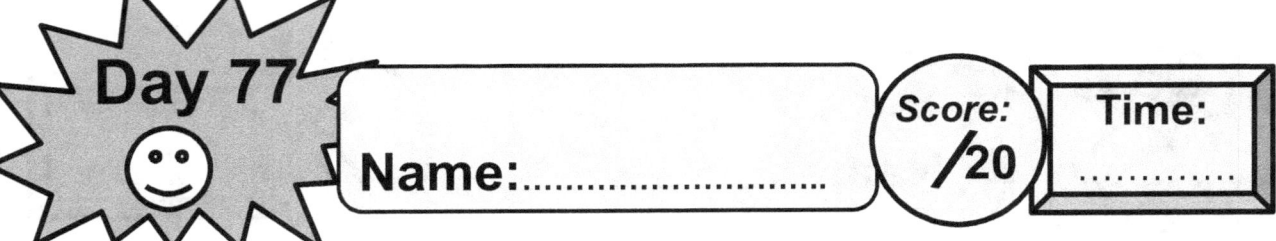

1	9	8	0	0	0	0
x 1	x 0	x 0	x 3	x 2	x 1	x 0

1	1	1	0	0	0	0
x 8	x 7	x 6	x 0	x 9	x 8	x 7

5	4	3	7	6	5	4
x 1	x 1	x 1	x 0	x 0	x 0	x 0

2	2	2	1	1	1	1
x 2	x 1	x 0	x 5	x 4	x 3	x 2

2	1	0	1	0	0	0
x 1	x 1	x 1	x 9	x 6	x 5	x 4

9	8	7	6	3	2	1
x 1	x 1	x 1	x 1	x 0	x 0	x 0

2	2	2	2	2	2	2
x 9	x 8	x 7	x 6	x 5	x 4	x 3

6	5	4	3	2	1	0
x 2	x 2	x 2	x 2	x 2	x 2	x 2

3	3	3	3	9	8	7
x 3	x 2	x 1	x 0	x 2	x 2	x 2

Name:..........................

Score: /20

Time:
....

2	3	1	4	3	3	0
x 4	x 4	x 6	x 7	x 8	x 0	x 3

1	2	3	4	5	6	7
x 1	x 3	x 3	x 3	x 3	x 3	x 3

8	9	4	4	2	4	4
x 3	x 9	x 0	x 1	x 2	x 3	x 4

4	9	4	4	7	0	1
x 5	x 6	x 7	x 0	x 9	x 4	x 4

2	3	4	5	6	7	8
x 4	x 4	x 4	x 1	x 6	x 4	x 0

7						
9	5	1	2	3	4	5
x 4	x 0	x 5	x 7	x 5	x 4	x 5

6	7	8	9	5	6	5
x 5	x 9	x 5	x 7	x 0	x 1	x 2

2	5	5	1	5	5	5
x 3	x 4	x 5	x 6	x 7	x 8	x 9

6	5	6	7	6	8	6
x 0	x 1	x 2	x 3	x 4	x 5	x 6

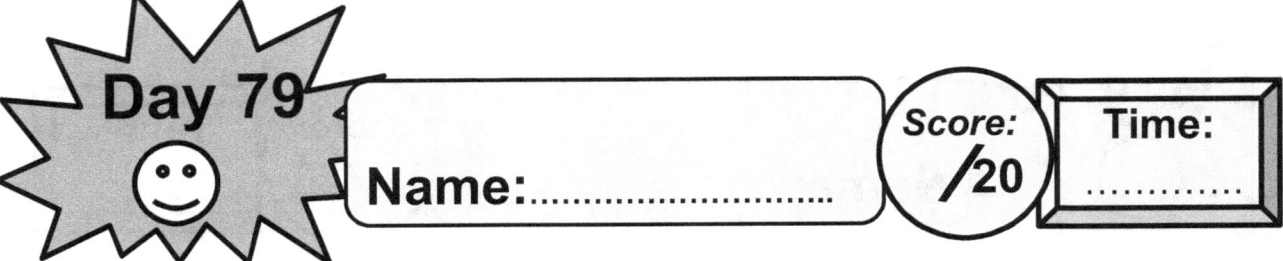

3 x 0	2 x 6	1 x 9	0 x 6	6 x 8	6 x 7	6 x 7
7 x 0	9 x 1	8 x 6	7 x 2	6 x 6	5 x 3	4 x 6
5 x 2	4 x 6	6 x 5	7 x 4	8 x 3	7 x 2	9 x 1
4 x 0	3 x 7	2 x 9	1 x 7	0 x 2	7 x 9	7 x 8
8 x 1	8 x 0	9 x 7	8 x 7	7 x 7	6 x 7	5 x 7
8 x 8	5 x 7	8 x 6	2 x 5	8 x 4	9 x 3	8 x 2
6 x 8	4 x 5	3 x 8	2 x 4	1 x 8	0 x 8	8 x 9
9 x 2	9 x 1	9 x 0	9 x 8	8 x 8	7 x 8	6 x 8
9 x 9	9 x 8	9 x 7	9 x 6	9 x 5	9 x 4	9 x 3

Day 80

Name:..............................

Score: /20

Time: ....

0 x 9	1 x 9	2 x 9	3 x 9	4 x 9	5 x 9	6 x 9
7 x 9	8 x 9	9 x 9	1 x 2	2 x 3	3 x 4	4 x 5
5 x 6	6 x 7	7 x 8	8 x 9	9 x 1	2 x 8	3 x 7
4 x 6	9 x 0	0 x 2	2 x 4	8 x 1	8 x 6	5 x 2
4 x 0	9 x 1	5 x 3	1 x 3	1 x 6	7 x 2	5 x 1
4 x 1	4 x 2	8 x 3	9 x 3	4 x 0	4 x 3	4 x 4
4 x 8	4 x 9	4 x 5	4 x 6	4 x 7	0 x 4	1 x 4
5 x 4	6 x 4	2 x 4	3 x 4	4 x 4	7 x 4	8 x 4
2 x 5	3 x 5	9 x 4	5 x 0	1 x 5	4 x 5	5 x 5

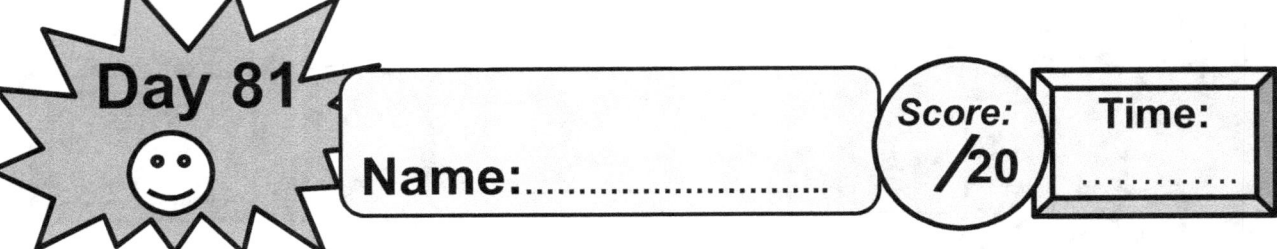

Score: /20   Time: ....

Name:..............................

0	0	0	0	0	0	0
x 6	x 5	x 4	x 3	x 2	x 1	x 0

3	2	1	0	0	0	0
x 0	x 0	x 0	x 0	x 9	x 8	x 7

1	9	8	7	6	5	4
x 1	x 0	x 0	x 0	x 0	x 0	x 0

1	1	1	1	1	1	1
x 8	x 7	x 6	x 5	x 4	x 3	x 2

5	4	3	2	1	0	1
x 1	x 1	x 1	x 1	x 1	x 1	x 9

2	2	2	9	8	7	6
x 2	x 1	x 0	x 1	x 1	x 1	x 1

2	2	2	2	2	2	2
x 9	x 8	x 7	x 6	x 5	x 4	x 3

6	5	4	3	2	1	0
x 2	x 2	x 2	x 2	x 2	x 2	x 2

3	3	3	3	9	8	7
x 3	x 2	x 1	x 0	x 2	x 2	x 2

Name:.............................

Score: /20

Time: ....

3	3	3	3	3	3	0
x 4	x 5	x 6	x 7	x 8	x 9	x 3

1	2	3	4	5	6	7
x 3	x 3	x 3	x 3	x 3	x 3	x 3

8	9	4	4	4	4	4
x 3	x 3	x 0	x 1	x 2	x 3	x 4

4	4	4	4	4	0	1
x 5	x 6	x 7	x 8	x 9	x 4	x 4

2	3	4	5	6	7	8
x 4	x 4	x 4	x 4	x 4	x 4	x 4

9	5	1	2	3	4	5
x 4	x 0	x 5	x 5	x 5	x 5	x 5

6	7	8	9	5	5	5
x 5	x 5	x 5	x 5	x 0	x 1	x 2

5	5	5	5	5	5	5
x 3	x 4	x 5	x 6	x 7	x 8	x 9

6	6	6	6	6	6	6
x 0	x 1	x 2	x 3	x 4	x 5	x 6

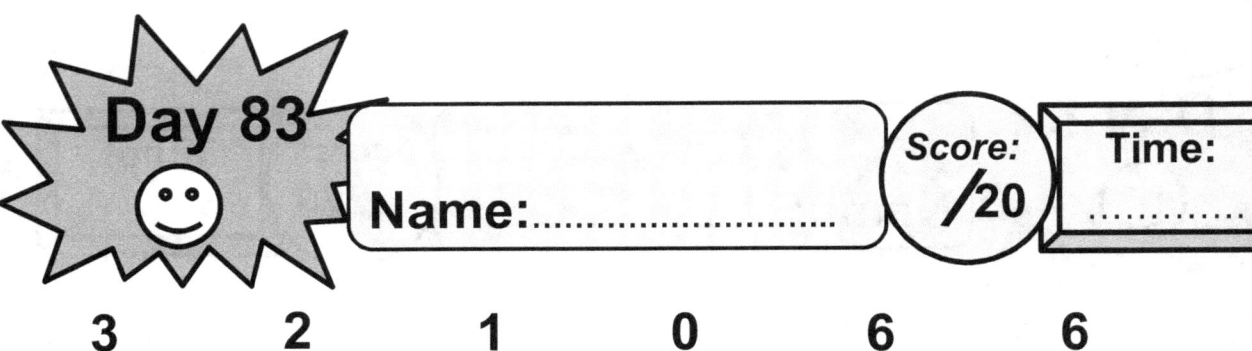

3 x 6	2 x 6	1 x 6	0 x 6	6 x 9	6 x 8	6 x 7
7 x 0	9 x 6	8 x 6	7 x 6	6 x 6	5 x 6	4 x 6
7 x 7	7 x 6	7 x 5	7 x 4	7 x 3	7 x 2	7 x 1
4 x 7	3 x 7	2 x 7	1 x 7	0 x 7	7 x 9	7 x 8
8 x 1	8 x 0	9 x 7	8 x 7	7 x 7	6 x 7	5 x 7
8 x 8	8 x 7	8 x 6	8 x 5	8 x 4	8 x 3	8 x 2
5 x 8	4 x 8	3 x 8	2 x 8	1 x 8	0 x 8	8 x 9
9 x 2	9 x 1	9 x 0	9 x 8	8 x 8	7 x 8	6 x 8
9 x 9	9 x 8	9 x 7	9 x 6	9 x 5	9 x 4	9 x 3

Day 84

Name:...........................

Score: /20

Time: ....

0	1	2	3	4	5	6
x 9	x 9	x 9	x 9	x 9	x 9	x 9

7	8	9	1	2	3	4
x 9	x 9	x 9	x 2	x 3	x 4	x 5

5	6	7	8	9	2	3
x 6	x 7	x 8	x 9	x 1	x 8	x 7

4	9	0	2	8	8	5
x 6	x 0	x 2	x 4	x 1	x 6	x 2

4	9	5	1	1	7	5
x 0	x 1	x 3	x 3	x 6	x 2	x 1

4	4	8	9	4	4	4
x 1	x 2	x 3	x 3	x 0	x 3	x 4

4	4	4	4	4	0	1
x 8	x 9	x 5	x 6	x 7	x 4	x 4

5	6	2	3	4	7	8
x 4	x 4	x 4	x 4	x 4	x 4	x 4

2	3	9	5	1	4	5
x 5	x 5	x 4	x 0	x 5	x 5	x 5

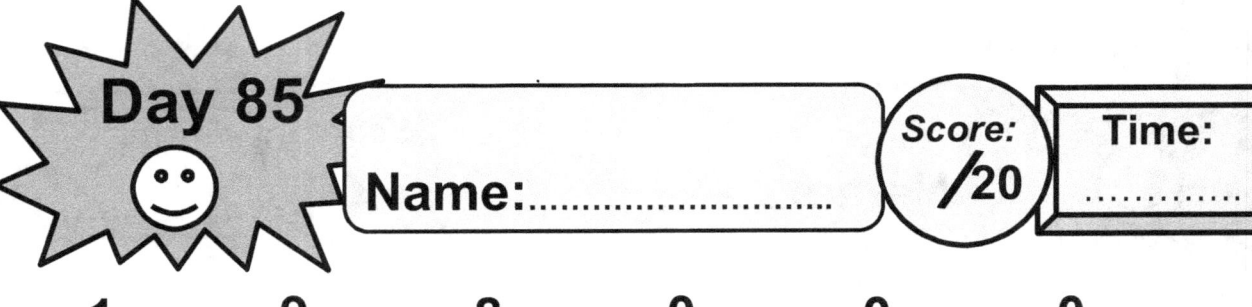

Name:...........................

Score: /20

Time: ....

1 x 1	9 x 0	8 x 0	0 x 3	0 x 2	0 x 1	0 x 0
1 x 8	1 x 7	1 x 6	0 x 0	0 x 9	0 x 8	0 x 7
5 x 1	4 x 1	3 x 1	7 x 0	6 x 0	5 x 0	4 x 0
2 x 2	2 x 1	2 x 0	1 x 5	1 x 4	1 x 3	1 x 2
2 x 1	1 x 1	0 x 1	1 x 9	0 x 6	0 x 5	0 x 4
9 x 1	8 x 1	7 x 1	6 x 1	3 x 0	2 x 0	1 x 0
2 x 9	2 x 8	2 x 7	2 x 6	2 x 5	2 x 4	2 x 3
6 x 2	5 x 2	4 x 2	3 x 2	2 x 2	1 x 2	0 x 2
3 x 3	3 x 2	3 x 1	3 x 0	9 x 2	8 x 2	7 x 2

Day 86

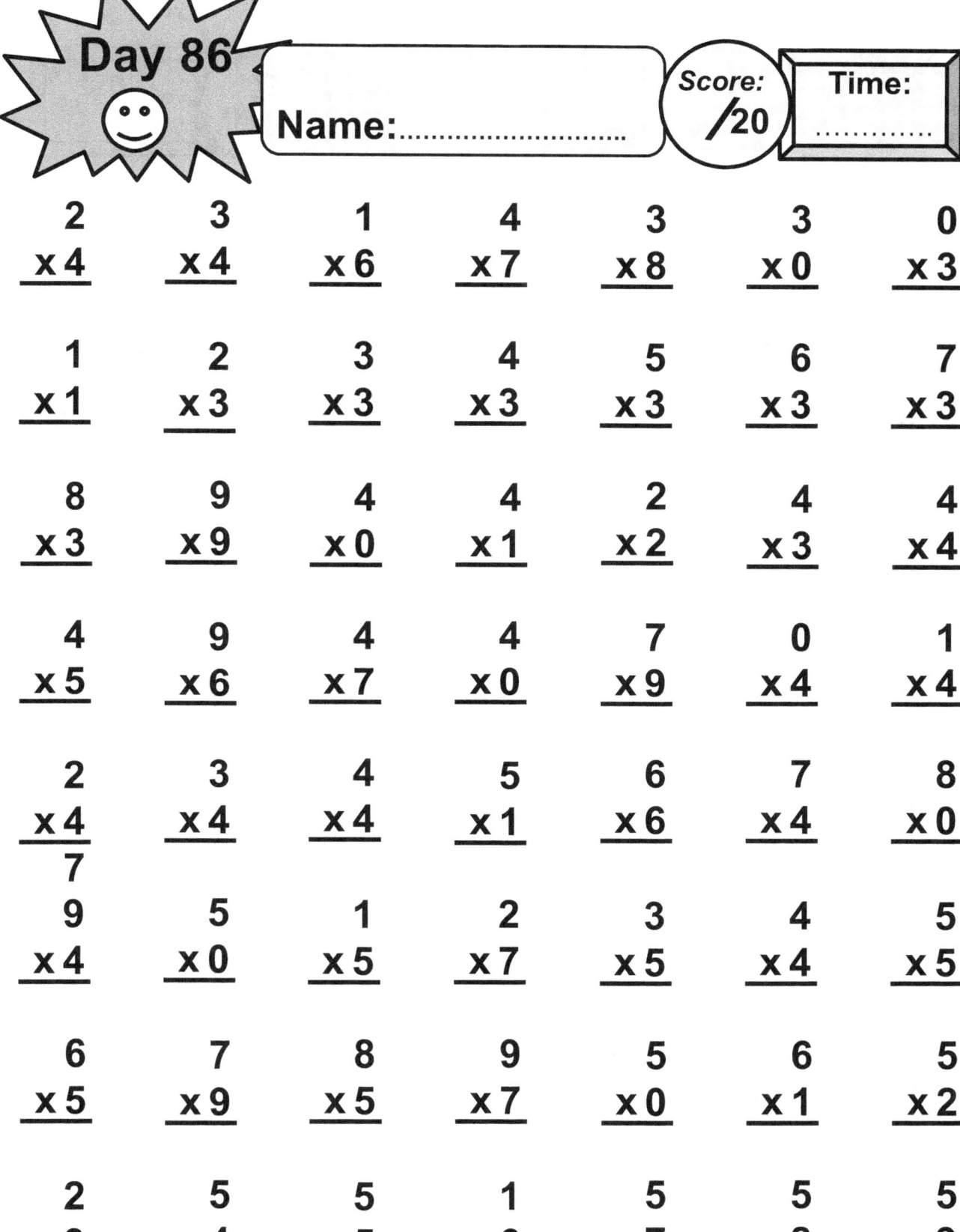

Name:..........................

Score: /20

Time: ....

2 x 4	3 x 4	1 x 6	4 x 7	3 x 8	3 x 0	0 x 3
1 x 1	2 x 3	3 x 3	4 x 3	5 x 3	6 x 3	7 x 3
8 x 3	9 x 9	4 x 0	4 x 1	2 x 2	4 x 3	4 x 4
4 x 5	9 x 6	4 x 7	4 x 0	7 x 9	0 x 4	1 x 4
2 x 4	3 x 4	4 x 4	5 x 1	6 x 6	7 x 4	8 x 0
7 9 x 4	5 x 0	1 x 5	2 x 7	3 x 5	4 x 4	5 x 5
6 x 5	7 x 9	8 x 5	9 x 7	5 x 0	6 x 1	5 x 2
2 x 3	5 x 4	5 x 5	1 x 6	5 x 7	5 x 8	5 x 9
6 x 0	5 x 1	6 x 2	7 x 3	6 x 4	8 x 5	6 x 6

3	2	1	0	6	6	6
x 0	x 6	x 9	x 6	x 8	x 7	x 7

7	9	8	7	6	5	4
x 0	x 1	x 6	x 2	x 6	x 3	x 6

5	4	6	7	8	7	9
x 2	x 6	x 5	x 4	x 3	x 2	x 1

4	3	2	1	0	7	7
x 0	x 7	x 9	x 7	x 2	x 9	x 8

8	8	9	8	7	6	5
x 1	x 0	x 7	x 7	x 7	x 7	x 7

8	5	8	2	8	9	8
x 8	x 7	x 6	x 5	x 4	x 3	x 2

6	4	3	2	1	0	8
x 8	x 5	x 8	x 4	x 8	x 8	x 9

9	9	9	9	8	7	6
x 2	x 1	x 0	x 8	x 8	x 8	x 8

9	9	9	9	9	9	9
x 9	x 8	x 7	x 6	x 5	x 4	x 3

Name:..........................

Score: /20

Time: ....

0	1	2	3	4	5	6
x 9	x 9	x 9	x 9	x 9	x 9	x 9

7	8	9	1	2	3	4
x 9	x 9	x 9	x 2	x 3	x 4	x 5

5	6	7	8	9	2	3
x 6	x 7	x 8	x 9	x 1	x 8	x 7

4	9	0	2	8	8	5
x 6	x 0	x 2	x 4	x 1	x 6	x 2

4	9	5	1	1	7	5
x 0	x 1	x 3	x 3	x 6	x 2	x 1

4	4	8	9	4	4	4
x 1	x 2	x 3	x 3	x 0	x 3	x 4

4	4	4	4	4	0	1
x 8	x 9	x 5	x 6	x 7	x 4	x 4

5	6	2	3	4	7	8
x 4	x 4	x 4	x 4	x 4	x 4	x 4

2	3	9	5	1	4	5
x 5	x 5	x 4	x 0	x 5	x 5	x 5

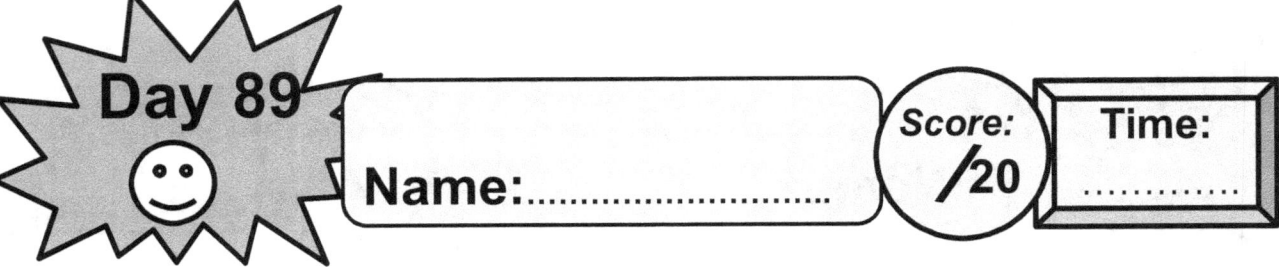

0	0	0	0	0	0	0
x 6	x 5	x 4	x 3	x 2	x 1	x 0

3	2	1	0	0	0	0
x 0	x 0	x 0	x 0	x 9	x 8	x 7

1	9	8	7	6	5	4
x 1	x 0	x 0	x 0	x 0	x 0	x 0

1	1	1	1	1	1	1
x 8	x 7	x 6	x 5	x 4	x 3	x 2

5	4	3	2	1	0	1
x 1	x 1	x 1	x 1	x 1	x 1	x 9

2	2	2	9	8	7	6
x 2	x 1	x 0	x 1	x 1	x 1	x 1

2	2	2	2	2	2	2
x 9	x 8	x 7	x 6	x 5	x 4	x 3

6	5	4	3	2	1	0
x 2	x 2	x 2	x 2	x 2	x 2	x 2

3	3	3	3	9	8	7
x 3	x 2	x 1	x 0	x 2	x 2	x 2

Day 90

Name:.........................

Score: /20

Time: ....

3	3	3	3	3	3	0
x 4	x 5	x 6	x 7	x 8	x 9	x 3

1	2	3	4	5	6	7
x 3	x 3	x 3	x 3	x 3	x 3	x 3

8	9	4	4	4	4	4
x 3	x 3	x 0	x 1	x 2	x 3	x 4

4	4	4	4	4	0	1
x 5	x 6	x 7	x 8	x 9	x 4	x 4

2	3	4	5	6	7	8
x 4	x 4	x 4	x 4	x 4	x 4	x 4

9	5	1	2	3	4	5
x 4	x 0	x 5	x 5	x 5	x 5	x 5

6	7	8	9	5	5	5
x 5	x 5	x 5	x 5	x 0	x 1	x 2

5	5	5	5	5	5	5
x 3	x 4	x 5	x 6	x 7	x 8	x 9

6	6	6	6	6	6	6
x 0	x 1	x 2	x 3	x 4	x 5	x 6

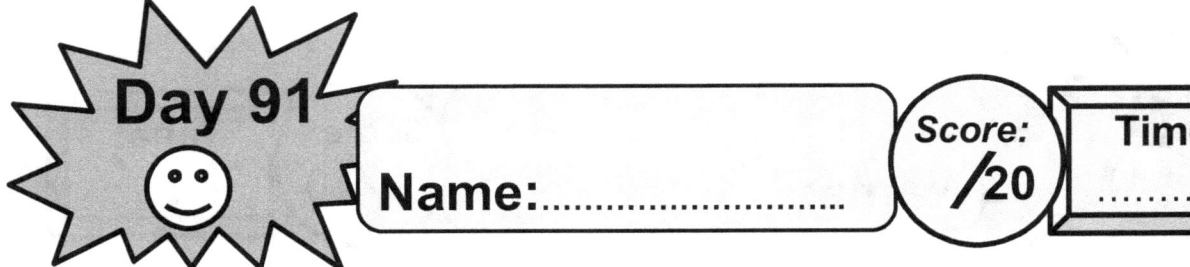

3 x 6	2 x 6	1 x 6	0 x 6	6 x 9	6 x 8	6 x 7
7 x 0	9 x 6	8 x 6	7 x 6	6 x 6	5 x 6	4 x 6
7 x 7	7 x 6	7 x 5	7 x 4	7 x 3	7 x 2	7 x 1
4 x 7	3 x 7	2 x 7	1 x 7	0 x 7	7 x 9	7 x 8
8 x 1	8 x 0	9 x 7	8 x 7	7 x 7	6 x 7	5 x 7
8 x 8	8 x 7	8 x 6	8 x 5	8 x 4	8 x 3	8 x 2
5 x 8	4 x 8	3 x 8	2 x 8	1 x 8	0 x 8	8 x 9
9 x 2	9 x 1	9 x 0	9 x 8	8 x 8	7 x 8	6 x 8
9 x 9	9 x 8	9 x 7	9 x 6	9 x 5	9 x 4	9 x 3

Day 92

Name:..............................

Score: /20

Time: ....

0	1	2	3	4	5	6
x 9	x 9	x 9	x 9	x 9	x 9	x 9

7	8	9	1	2	3	4
x 9	x 9	x 9	x 2	x 3	x 4	x 5

5	6	7	8	9	2	3
x 6	x 7	x 8	x 9	x 1	x 8	x 7

4	9	0	2	8	8	5
x 6	x 0	x 2	x 4	x 1	x 6	x 2

4	9	5	1	1	7	5
x 0	x 1	x 3	x 3	x 6	x 2	x 1

4	4	8	9	4	4	4
x 1	x 2	x 3	x 3	x 0	x 3	x 4

4	4	4	4	4	0	1
x 8	x 9	x 5	x 6	x 7	x 4	x 4

5	6	2	3	4	7	8
x 4	x 4	x 4	x 4	x 4	x 4	x 4

2	3	9	5	1	4	5
x 5	x 5	x 4	x 0	x 5	x 5	x 5

Name:........................

Score: /20

Time: ....

1	9	8	0	0	0	0
x1	x0	x0	x3	x2	x1	x0

1	1	1	0	0	0	0
x8	x7	x6	x0	x9	x8	x7

5	4	3	7	6	5	4
x1	x1	x1	x0	x0	x0	x0

2	2	2	1	1	1	1
x2	x1	x0	x5	x4	x3	x2

2	1	0	1	0	0	0
x1	x1	x1	x9	x6	x5	x4

9	8	7	6	3	2	1
x1	x1	x1	x1	x0	x0	x0

2	2	2	2	2	2	2
x9	x8	x7	x6	x5	x4	x3

6	5	4	3	2	1	0
x2	x2	x2	x2	x2	x2	x2

3	3	3	3	9	8	7
x3	x2	x1	x0	x2	x2	x2

Day 94

Name:...........................

Score: /20

Time: ....

2 x 4	3 x 4	1 x 6	4 x 7	3 x 8	3 x 0	0 x 3
1 x 1	2 x 3	3 x 3	4 x 3	5 x 3	6 x 3	7 x 3
8 x 3	9 x 9	4 x 0	4 x 1	2 x 2	4 x 3	4 x 4
4 x 5	9 x 6	4 x 7	4 x 0	7 x 9	0 x 4	1 x 4
2 x 4	3 x 4	4 x 4	5 x 1	6 x 6	7 x 4	8 x 0
7 9 x 4	5 x 0	1 x 5	2 x 7	3 x 5	4 x 4	5 x 5
6 x 5	7 x 9	8 x 5	9 x 7	5 x 0	6 x 1	5 x 2
2 x 3	5 x 4	5 x 5	1 x 6	5 x 7	5 x 8	5 x 9
6 x 0	5 x 1	6 x 2	7 x 3	6 x 4	8 x 5	6 x 6

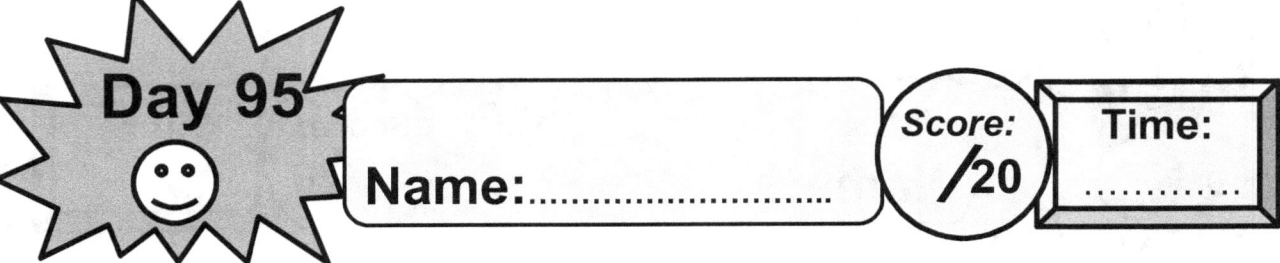

3 x 0	2 x 6	1 x 9	0 x 6	6 x 8	6 x 7	6 x 7
7 x 0	9 x 1	8 x 6	7 x 2	6 x 6	5 x 3	4 x 6
5 x 2	4 x 6	6 x 5	7 x 4	8 x 3	7 x 2	9 x 1
4 x 0	3 x 7	2 x 9	1 x 7	0 x 2	7 x 9	7 x 8
8 x 1	8 x 0	9 x 7	8 x 7	7 x 7	6 x 7	5 x 7
8 x 8	5 x 7	8 x 6	2 x 5	8 x 4	9 x 3	8 x 2
6 x 8	4 x 5	3 x 8	2 x 4	1 x 8	0 x 8	8 x 9
9 x 2	9 x 1	9 x 0	9 x 8	8 x 8	7 x 8	6 x 8
9 x 9	9 x 8	9 x 7	9 x 6	9 x 5	9 x 4	9 x 3

Score: /20

Time:

Name:

0	1	2	3	4	5	6
x 9	x 9	x 9	x 9	x 9	x 9	x 9

7	8	9	1	2	3	4
x 9	x 9	x 9	x 2	x 3	x 4	x 5

5	6	7	8	9	2	3
x 6	x 7	x 8	x 9	x 1	x 8	x 7

4	9	0	2	8	8	5
x 6	x 0	x 2	x 4	x 1	x 6	x 2

4	9	5	1	1	7	5
x 0	x 1	x 3	x 3	x 6	x 2	x 1

4	4	8	9	4	4	4
x 1	x 2	x 3	x 3	x 0	x 3	x 4

4	4	4	4	4	0	1
x 8	x 9	x 5	x 6	x 7	x 4	x 4

5	6	2	3	4	7	8
x 4	x 4	x 4	x 4	x 4	x 4	x 4

2	3	9	5	1	4	5
x 5	x 5	x 4	x 0	x 5	x 5	x 5

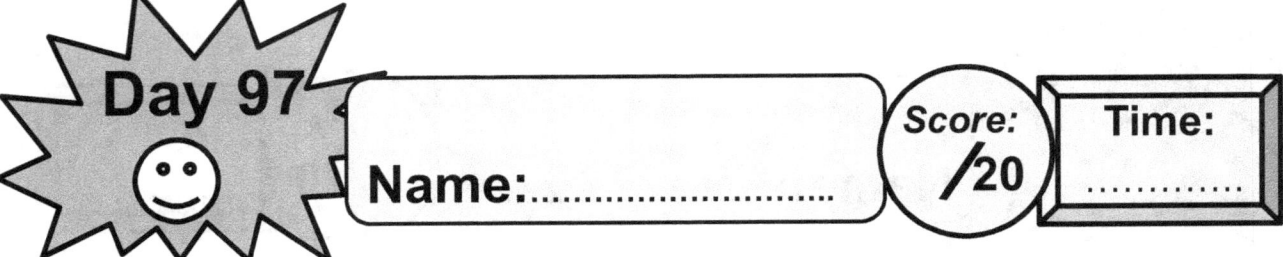

Name:............................

Score: /20

Time: ....

0	0	0	0	0	0	0
x 6	x 5	x 4	x 3	x 2	x 1	x 0

3	2	1	0	0	0	0
x 0	x 0	x 0	x 0	x 9	x 8	x 7

1	9	8	7	6	5	4
x 1	x 0	x 0	x 0	x 0	x 0	x 0

1	1	1	1	1	1	1
x 8	x 7	x 6	x 5	x 4	x 3	x 2

5	4	3	2	1	0	1
x 1	x 1	x 1	x 1	x 1	x 1	x 9

2	2	2	9	8	7	6
x 2	x 1	x 0	x 1	x 1	x 1	x 1

2	2	2	2	2	2	2
x 9	x 8	x 7	x 6	x 5	x 4	x 3

6	5	4	3	2	1	0
x 2	x 2	x 2	x 2	x 2	x 2	x 2

3	3	3	3	9	8	7
x 3	x 2	x 1	x 0	x 2	x 2	x 2

Name:............................

Score: /20

Time: ....

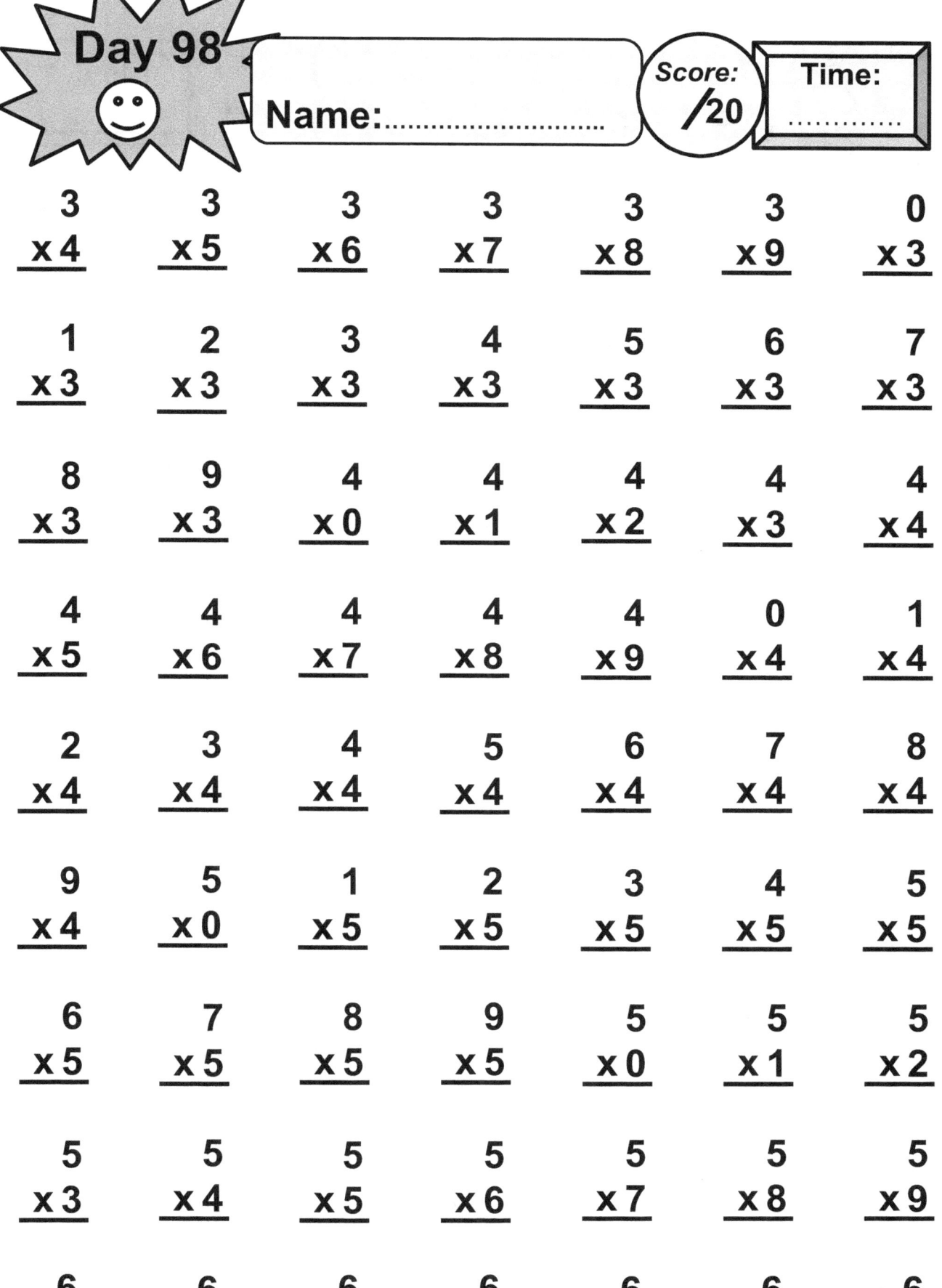

3	3	3	3	3	3	0
x 4	x 5	x 6	x 7	x 8	x 9	x 3

1	2	3	4	5	6	7
x 3	x 3	x 3	x 3	x 3	x 3	x 3

8	9	4	4	4	4	4
x 3	x 3	x 0	x 1	x 2	x 3	x 4

4	4	4	4	4	0	1
x 5	x 6	x 7	x 8	x 9	x 4	x 4

2	3	4	5	6	7	8
x 4	x 4	x 4	x 4	x 4	x 4	x 4

9	5	1	2	3	4	5
x 4	x 0	x 5	x 5	x 5	x 5	x 5

6	7	8	9	5	5	5
x 5	x 5	x 5	x 5	x 0	x 1	x 2

5	5	5	5	5	5	5
x 3	x 4	x 5	x 6	x 7	x 8	x 9

6	6	6	6	6	6	6
x 0	x 1	x 2	x 3	x 4	x 5	x 6

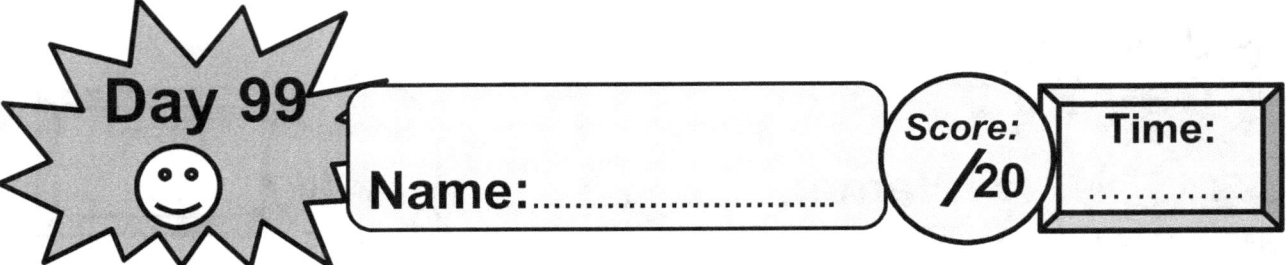

3 x 6	2 x 6	1 x 6	0 x 6	6 x 9	6 x 8	6 x 7
7 x 0	9 x 6	8 x 6	7 x 6	6 x 6	5 x 6	4 x 6
7 x 7	7 x 6	7 x 5	7 x 4	7 x 3	7 x 2	7 x 1
4 x 7	3 x 7	2 x 7	1 x 7	0 x 7	7 x 9	7 x 8
8 x 1	8 x 0	9 x 7	8 x 7	7 x 7	6 x 7	5 x 7
8 x 8	8 x 7	8 x 6	8 x 5	8 x 4	8 x 3	8 x 2
5 x 8	4 x 8	3 x 8	2 x 8	1 x 8	0 x 8	8 x 9
9 x 2	9 x 1	9 x 0	9 x 8	8 x 8	7 x 8	6 x 8
9 x 9	9 x 8	9 x 7	9 x 6	9 x 5	9 x 4	9 x 3

Day 100

Name:...........................

Score: /20

Time: ....

0	1	2	3	4	5	6
x9	x9	x9	x9	x9	x9	x9

7	8	9	1	2	3	4
x9	x9	x9	x2	x3	x4	x5

5	6	7	8	9	2	3
x6	x7	x8	x9	x1	x8	x7

4	9	0	2	8	8	5
x6	x0	x2	x4	x1	x6	x2

4	9	5	1	1	7	5
x0	x1	x3	x3	x6	x2	x1

4	4	8	9	4	4	4
x1	x2	x3	x3	x0	x3	x4

4	4	4	4	4	0	1
x8	x9	x5	x6	x7	x4	x4

5	6	2	3	4	7	8
x4	x4	x4	x4	x4	x4	x4

2	3	9	5	1	4	5
x5	x5	x4	x0	x5	x5	x5

Day 101

Name:..........................

Score: /20

Time: ....

1 x 1	9 x 0	8 x 0	0 x 3	0 x 2	0 x 1	0 x 0
1 x 8	1 x 7	1 x 6	0 x 0	0 x 9	0 x 8	0 x 7
5 x 1	4 x 1	3 x 1	7 x 0	6 x 0	5 x 0	4 x 0
2 x 2	2 x 1	2 x 0	1 x 5	1 x 4	1 x 3	1 x 2
2 x 1	1 x 1	0 x 1	1 x 9	0 x 6	0 x 5	0 x 4
9 x 1	8 x 1	7 x 1	6 x 1	3 x 0	2 x 0	1 x 0
2 x 9	2 x 8	2 x 7	2 x 6	2 x 5	2 x 4	2 x 3
6 x 2	5 x 2	4 x 2	3 x 2	2 x 2	1 x 2	0 x 2
3 x 3	3 x 2	3 x 1	3 x 0	9 x 2	8 x 2	7 x 2

2 x 4	3 x 4	1 x 6	4 x 7	3 x 8	3 x 0	0 x 3
1 x 1	2 x 3	3 x 3	4 x 3	5 x 3	6 x 3	7 x 3
8 x 3	9 x 9	4 x 0	4 x 1	2 x 2	4 x 3	4 x 4
4 x 5	9 x 6	4 x 7	4 x 0	7 x 9	0 x 4	1 x 4
2 x 4	3 x 4	4 x 4	5 x 1	6 x 6	7 x 4	8 x 0
7 9 x 4	5 x 0	1 x 5	2 x 7	3 x 5	4 x 4	5 x 5
6 x 5	7 x 9	8 x 5	9 x 7	5 x 0	6 x 1	5 x 2
2 x 3	5 x 4	5 x 5	1 x 6	5 x 7	5 x 8	5 x 9
6 x 0	5 x 1	6 x 2	7 x 3	6 x 4	8 x 5	6 x 6

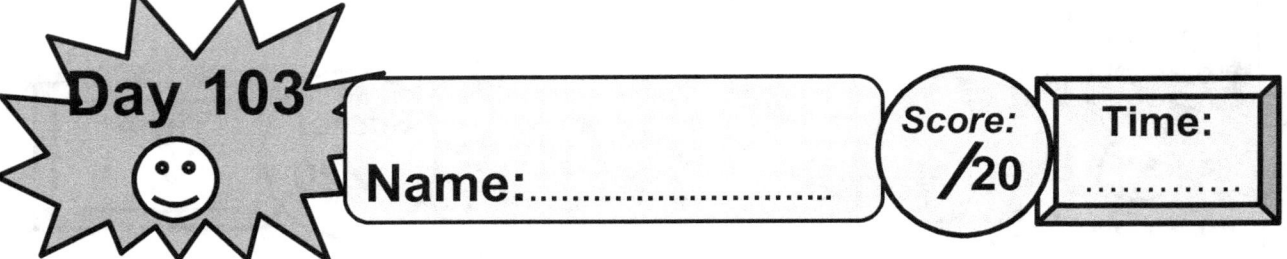

3	2	1	0	6	6	6
x 0	x 6	x 9	x 6	x 8	x 7	x 7

7	9	8	7	6	5	4
x 0	x 1	x 6	x 2	x 6	x 3	x 6

5	4	6	7	8	7	9
x 2	x 6	x 5	x 4	x 3	x 2	x 1

4	3	2	1	0	7	7
x 0	x 7	x 9	x 7	x 2	x 9	x 8

8	8	9	8	7	6	5
x 1	x 0	x 7	x 7	x 7	x 7	x 7

8	5	8	2	8	9	8
x 8	x 7	x 6	x 5	x 4	x 3	x 2

6	4	3	2	1	0	8
x 8	x 5	x 8	x 4	x 8	x 8	x 9

9	9	9	9	8	7	6
x 2	x 1	x 0	x 8	x 8	x 8	x 8

9	9	9	9	9	9	9
x 9	x 8	x 7	x 6	x 5	x 4	x 3